Sparing Nature

Sparing
Nature

The Conflict between Human Population
Growth and Earth's Biodiversity

Jeffrey K. McKee

 Rutgers University Press
New Brunswick, New Jersey, and London

First paperback printing, 2005

Library of Congress Cataloging-in-Publication Data

McKee, Jeffrey Kevin
  Sparing nature : the conflict between human population growth and earth's biodiversity /
Jeffrey K. McKee.
        p. cm.
  Includes bibliographic references and index.
  ISBN 0-8135-3141-1 (cloth : alk. paper) ISBN 0-8135-3558-1 (pbk. : alk. paper)
    1.   Population—Environmental aspects. 2. Population. 3. Biological diversity. I. Title.

  HB849.415.M38 2003
  333.95—dc21   2002070506

British Cataloging-in-Publication data for this book is available from the British Library.

Design by John Romer

Manufactured in the United States of America

This one is for Nathan and Aaron,
with a special thanks to Jean

Contents

Preface

After finally ushering my last book into press in 2000, I soon felt another book formulating itself in my mind. I enjoyed writing *The Riddled Chain*, although the topic of its final chapter — the relevance of evolutionary research to contemporary environmental concerns — was not covered as fully as I would have liked. But there were so many good books on humans and the current extinction crisis that I thought such a project would not be worth my while. Then I discovered a "missing link," albeit not of the kind paleoanthropologists usually like to find. This one was the gap in the literature regarding any connection between human population growth and biodiversity losses.

The discovery came while I was teaching a course called Human Ecological Adaptations, covering the environmental context of our lineage over the past few million years. As we covered the past ten thousand years, which we refer to as the "present," the course focused on two major trends: the explosive growth of the human population and the ongoing mass extinction of plants and animals. My first surprise was that a frightening number of my students were unaware of either trend, despite constant exposure to the varied media that explore them in depth. So perhaps one more book would help send out the message.

Nevertheless, I've always taught that the two global trends are related. A number of my students wanted to research this relationship further for their term papers, and I used to instruct them that there was a wealth of literature on the topic. As it turned out, I was wrong. There are many books on the biodiversity crisis, but at best they touch on human population growth for a few paragraphs. Likewise, despite a wealth of literature on human demography, its relationship to the extinction of other life-forms has barely been mentioned.

At first, the lack of literature led me to doubt the connection between the paired trends, so I began reading more and researched a few ideas on my own. It became abundantly clear that the connection is real, albeit relatively unnoticed. Thus my goal was set: to fill the gap in the literature with an extended essay, synthesizing the best of modern research and explaining it with a style that could awaken students young and old. The subject was too important to leave untreated.

The challenge was to tackle such a large and complex topic without getting the book mired in unreadable detail. In my last book I tried to overcome a similar problem through the use of anecdotes, analogies, and occasional light relief. For this new book the anecdotes would be easy, as I have been privileged to visit many parts of the world from which I could draw examples — including my own backyard. Yet as for light relief, some of my friends asked me how I could possibly find humor in the overpopulation quagmire and the biodiversity crisis. So in writing on such a serious topic, I recalled the comment of a student who took my Human Ecological Adaptations course, from which this book got its inspiration. She said, to my amusement, "This was the most depressing course I've ever enjoyed." From that springboard I took in other books that take a lighter approach to the maladies of today: comic novelist Douglas Adams's *Last Chance to See*, cartoonist Gary Larson's *There's a Hair in My Dirt*, and my favorite, *The Lorax*, by Theodor Geisel (a k a Dr. Seuss).

Nevertheless, the two main messages of this book are quite serious: (1) There is a connection between human population growth and biodiversity loss that goes back over a million years, and that becomes particularly evident in the past ten thousand years. (2) The greatest and most effective conservation measure to save earth's biodiversity is to halt the growth of the human population, and perhaps reduce our numbers. As I combed the academic research journals, I was amazed at the amount of supportive evidence; nearly two hundred scientific works are cited in this book, and that is just a sampling of what is out there. My apologies to those scholars who have done relevant work but did not get cited — the enormity of the topic meant I had to pick and choose research that kept to the focus of the book.

One cannot be an expert in all realms of science, so I had some help from a number of people in straightening out details and getting rid of minor errors that could have detracted from the message of *Sparing Nature*. Any remaining mistakes are entirely my own. I'm also in debt to a number of readers who spotted the many typographical errors in the draft manuscript, and who gently persuaded me to change some of my more tortuous sentences into readable ones. In alphabetical order, my team of helpers were David Fooce, Kris Gremillion, Jerry Harris, Scott McGraw, Bill McKee, Steve McKee, and Steve Rissing. An anonymous referee also provided valuable insights and encouragement. I thank them all most sincerely. I am also very grateful to Michael Masters and Tim McKee for their artistry in giving this book some delightful imagery.

The crew at Rutgers University Press has always been a delight to work with, and makes the publishing process seem smooth and easy. I thank editors Helen Hsu, Audra Wolfe, Marilyn Campbell, Amy Rashap, and a host of other hardworking souls. My insightful copy editor, Will Hively, also did a meticulous job on the final version of the text, for which I am most grateful.

And, of course, none of this would have been possible without the daily love and support of Aaron, Nathan, and Jean.

Sparing Nature

1

CHAPTER

Sparing Nature

Through the animal and vegetable kingdoms Nature has scattered the seeds of life abroad with the most profuse and liberal hand; but has been comparatively sparing in the room and the nourishment necessary to rear them. The germs of existence contained in this earth, if they could freely develop themselves, would fill millions of worlds in the course of a few thousand years. Necessity, that imperious, all-pervading law of nature, restrains them within the prescribed bounds. The race of plants and the race of animals shrink under this great restrictive law; and man cannot by any efforts of reason escape from it.

THOMAS ROBERT MALTHUS
An Essay on the Principle of Population, 1798

ALONG a scenic stretch of the Olentangy River, in central Ohio, is a rock jutting out of the water where I like to sit and think. I often refer to this rock and its surroundings as my "office." As the water rushes past me I can enjoy observing a wide variety of plants and animals while I cogitate about life's origins and destiny. Where there is water there is life, and within view is a diverse array of life, or rich biodiversity.

Certain of the diverse biological organisms at the office are among my favorites. The damselflies seem to perform an aerial dance for me above the calmer recesses near the shore, while water striders skate on the clear surface below. Occasionally a fish jumps up from the middle of the river and then quickly retreats to the depths, while other creatures find sustenance in the riffles where the shallow water rushes over the pebbles. The children who often visit my office with the local naturalists are always astounded by the living treasures their nets catch in the riffles: mussels and darters adapted to the fast-moving water.

In the shade of an old and large sycamore tree, I glimpse a wealth of birds landing and taking off from its branches, flying on to a locust tree on the opposite bank. And, if I'm lucky, I'll get to see a passing deer or woodchuck.

Other representatives of life's diverse array are much less noticeable, but no less intriguing. Lying underneath the fallen sycamore leaves on the banks are all sorts of tiny organisms that enrich the soil. Many of them I cannot see, for there is no microscope at my office, but I know they are there. And they do a good job of cleaning up after the leafy "paperwork" that comes across my "desk" with the breeze.

Although my office is usually pleasant, sometimes the accommodations are not very practical. In the summer the insects can be a nuisance, and often, quite literally, I get a bug in my laptop computer. In the winter it can be bitterly cold, and with the return of warmer weather, the melted snow and spring rains conspire to submerge my rock. Despite the difficulties of accessing my office in the spring, the raised water level does have its advantages. The sound of the rushing water at that time of year is sufficient to drown out the jarring noises from the trucks and cars on the nearby "scenic highway." And I can view a different array of creatures and life stages that comprise the river's life system each spring.

Even in the summer and fall, when I utilize the office the most, there is really no room for my files and books. Like many people, I need a lot of space for the tools of my trade, but at this office my needs must be met by what is in my laptop computer or in my head. Moreover, my work is often interrupted, for I have to share the space with others who also come to the river to escape the clutter of suburban life. I'm happy to share the office with them, unless I see their picnic garbage floating down the river.

Farther downstream, and a short drive down the accompanying highway, is my real office at The Ohio State University. There, within the city of Columbus, the river takes on a vastly different nature. There are few trees to provide shade, and mammalian life is largely restricted to students and squirrels (not to mention the bats inhabiting the antiquated building that houses my Department of Anthropology). In place of the rich biodiversity are classroom buildings, libraries, and an imposing stadium that emits loud roars during Saturday afternoon football games in the fall. It is fun being among the hundred thousand or so people who create the cacophony in the stadium — indeed, I love all the trappings of the university and the lifestyle it provides me. But if I want to study biodiversity along the university's stretch of the river, then I have to find life's riches primarily in books and laboratories.

People need space and resources, as do all living beings. But nature "has been comparatively sparing in the room and the nourishment necessary to rear them" indeed. That humans and other organisms often compete for these commodities is the subject of this book.

Life at Home

I have yet another office taking up space at home—sometimes I need a retreat from the insects and students that bug me at my other offices. It seems humans can always find a way to use space when it is available, one way or another. My house in suburbia is on recently converted agricultural land. One can easily tell by the lack of tall trees in the area. Fields of corn and wheat have been permanently replaced by stretches of grass and aluminum-sided human shelters, divided up by an intricate matrix of asphalt roads.

Just to the north, reaching farther and farther from the city, are more homes under construction, taking up more and more of the valuable fields of grain. The city population is growing, and growing rapidly. This is natural. Humans love to reproduce—babies bring us much joy, and the process of making them is a great source of pleasure too. So the population grows. And, along with the new homes to house this population are new office buildings and new malls, as well as new roads and new parking lots to ease our way around the businesses—all taking up space, and covering what was once rich farmland.

Some people take a look at the loss of farmland and get worried about where our food is going to be grown. If there is less farmland to feed more and more people, ultimately the day will come when there won't be enough food to go around. This may or may not be a legitimate concern for now; indeed the amount of farmland has stayed relatively constant in the United States despite suburban sprawl, and new patches of land around the world are being cultivated with ever more sophisticated and efficient techniques. True, if the trend continues, then ultimately that day of concern *will* come, and we will have to reckon with the consequences. But for right now it seems that, technically speaking, there should be enough food and land—for humans, that is.[1] One should note, however, that famine is still quite common in many human populations. And one should wonder why.

What concerns me more than the loss of agricultural land is the demise of what existed there *before* it was cleared for crops and livestock. The forests that once covered Ohio are now all but gone, save a few patches, and the rich soil they created is being depleted by careless farming practices or getting covered with asphalt. With the natural forests and prairies went the plants and animals that once lived there. Also gone are the swamps and bogs that gave refuge to an amazing assortment of creatures. Nature's land became our land—we subsumed nature for our own purposes. Our growing population and its agricultural heritage take up the space and resources once occupied by a richer diversity of life — a greater biodiversity. The consequent demise of this biodiversity through extinctions great and small is a problem, and a big one at that.

Before we worry too much about this problem of extinction, we need to verify that it is indeed true that something is awry in the living world. Legions of scientists are out there talking about a biodiversity "crisis" and a mass extinction today on par with those of the past that decimated, for example, most dinosaur populations. Does human population growth, as well as the actions of people in general, really lead to biodiversity loss? Are human beings taking too large a share of earth's resources under the "great restrictive law" of nature? Will this place our planet's health in jeopardy? It is our nature, and indeed our duty, to question the scientific basis of the alleged crisis, for our conclusions will shape the policies and actions that we must pursue in the future.

Let me suggest a simple exercise so that you can test for yourself the dire warnings of the scientific community. Take some time to walk around the land on which you live, and count the number of different kinds of living organisms you find. You need not be a proficient naturalist to do this, for I am suggesting more of a thought experiment than rigorous scientific research. The count may be more time-consuming if you have an extensive plot of land with a well-tended garden than if you live in an apartment block in the middle of a city. But once you get a reasonable idea of life's richness at home, go to a comparable area in some undisturbed wild tract near your home (if you can find one) and count the life-forms again. Chances are very good that you will find more plant and animal species in the wild than at home. In other words, natural areas will tend to have greater biodiversity.

I've tried this exercise at my home and in a patch of woods near my Olentangy "office." My academic specialty deals with fossils, not living beings, so I was no more proficient at this exercise than most suburbanites. And, in order to make the exercise simpler, I dealt only with easily visible plant and animal species. Despite my fascination with various life-forms that grow in my shower stall or on last week's bread, I wasn't quite sure how to identify comparable things that grow in the woods. Likewise, I did not count the yeasts that I cultivate to make my bread or my home-brewed beer, for that adds more complications. But simple thought experiments can be quite revealing, even to an amateur.

The revelations of my count are portrayed in the accompanying illustration (fig. 1.1). Let's start with the mammals. I relish the visits of squirrels and rabbits to my yard, as do my two young sons. I even have a squirrel feeder to attract the little critters (it attracts birds as well). But I do not have flying squirrels emerging from the trees at night, or even fox squirrels so common in the woods — just gray squirrels and chipmunks. No woodchuck or deer has ever wandered by, having found the matrix of roads impenetrable. So the woods contain richer mammalian biodiversity.

It is worth noting that there are more deer in Ohio now than there were a hundred years ago. Doesn't that run counter to my observation of reduced biodiversity in the presence of humans? Not really, for diversity is measured by the sum of all species, not the population size of one. Many of those species are missing, including some of the larger predators that could take a deer. Although I was amazed to learn of foxes and coyotes still in our woods, we

mammals

plants

birds

insects

FIGURE 1.1 The number of species tends to go up in natural areas versus residential ones. These drawings represent the relative proportions of biodiversity at my home (*white*) and in a comparable-size patch of Ohio woods (*black*).

have squeezed out wolves and the lot of large predators from most of the state, and their prey have rebounded. That, along with other products of human intervention, has led to such success among deer populations that the local park rangers have to cull them on an annual basis. Leaving too many deer in the parks would quickly decimate the limited vegetative food sources. Deer overpopulation is as bad as human overpopulation, so we are forced to take nature management into our own hands. There may be more deer now, and more people, but there has been a net loss of mammalian biodiversity as a result of constructing our network of homes.

Despite my squirrel and bird feeder, the birds I see around my home are also less diverse than those found in the woods. This is due partly to the lack of tall trees on this former agricultural land. My family and I enjoy the cardinals, our state bird, that frequent the feeder, and my boys can identify most of the birds we get. But in the woods their knowledge of nature is comparatively limited, and they see an array of birds that baffle their young minds. Mine too: within the first hour of my experiment in the woods, I saw more bird species than I had seen during the entire year at home.

On the other hand, the number of plants at my house came remarkably close to that found in the wild patch I chose. But we are avid gardeners and plant many varied species to make our home more attractive. More attractive to *us*, that is. The diversity fails to lure or feed many mammals, birds, or reptiles. Indeed, we have no reptiles (nor will we, until one of my sons asks for a pet chameleon). Even the omnipresent insects shortchange us due to the neatly kept nature of our garden.

I should mention that many of the plant species decorating our lawn are not native to Ohio, and that many would not survive without our constant attention. Just ignore your lawn for one or two years, and there won't be much left of its pampered grasses. And the trees and bushes stand little chance of reproducing, for they are too far removed from other members of their species for pollination to stand a chance. But diversity is diversity, so we must count them in this experiment. It is worth noting, however, that I chose a patch in the woods with particularly tall trees. Their leafy canopy catches one of the greatest resources: energy from sunlight. This in turn limits the number of plants that can catch the rays on the woodland floor and that in turn might feed the deer. All life-forms compete for space and other resources.

I should also admit that I actively reduce plant biodiversity at my house. Not only did I cut down the sole representative of the crab apple tree species, but I must confess to occasionally going on an herbicidal rampage. I kill unwanted weeds such as the dandelion, and fight battles against the constant onslaught of opportunistic biological invaders. What constitutes a "weed" is in the eye of the beholder, and the standards of suburban lawn care dictate that certain plants are not desirable in the neighborhood (despite the fact that dandelions are a nutritious food). Funny, though, despite the dandelions' aggressive stance at home, I saw very few of them in the woods—and there usually near a footpath.

Counting insects is a bit more difficult than one might imagine. Some did me the favor of flying right into my face or biting me, thereby letting their presence be known. Others hid under rocks or in rotting logs. But with persistence one can find many different types. One need not necessarily identify them for this experiment: my notes have "little orange hopping bug" and similar nonscientific names. On the other hand, one must be careful not to raise the numbers artificially by counting bugs in their various developmental phases—you can't count a caterpillar and a butterfly. Keep at it, though. The numbers can be astounding because insect biodiversity is so great.

Okay, I've also committed insecticide. At home I kill bugs throughout the year, and feel little or no remorse. Yet insects are still the most diverse group of creatures at my house. Indeed, more than half of all living species are insects, and within each species the number of individuals can dwarf the human population size. For the most part this is good, for we could not live without them—insects do a lot of work for us such as pollinating plants, feeding birds, eating other insects, and more. We just don't want them in our homes, and sometimes need to eliminate them from our gardens. Life is more comfortable in the absence of wasps, cockroaches, and termites, though I was pleased and surprised to see a honeybee in my backyard after a conspicuous absence. By the way, I recommend that you take the time to identify at least some of the insects for this thought experiment. I learned what a termite looked like just in time to preserve the structure and economic value of my house!

Out in the woods insects can also be deemed undesirable. I can't tell you how many mosquito bites I had to endure to complete my experiment. But it was worth it. The woods harbor an amazing diversity of insects. And I counted only the kinds I could see from the ground. Up in the tall trees there must be even more bugs, doing the work of nature. I can hear many up there, such as the loud cicadas. Had I dug into the ground or dissected deeper into decaying logs, still more insects would have been found. Despite this obvious undercount, the insects of the woods easily exceeded those at home in biodiversity.

Most people who try this thought experiment will come out with similar results. Or the results may be worse. Those living in the "concrete jungle" of the city may find very little biological diversity around them. Perhaps, however, you tried this experiment at home and came out with a different result. If so, congratulations on the diversity you harbor. On the other hand, chances are good that many of the diverse organisms, particularly your house and garden plants, are not native. Or you may live in a desert environment and find more diversity of life around your house than beyond the borders of the city. In defense of suburban sprawl, syndicated columnist George Will once wrote that "lawns replace desert at the rate of about an acre an hour. Is that alarming?"[2] In some way it is, for those lawns do not come for free. Think of the water that you utilize to grow and nurture your diversity, and the energy — another key resource — it takes to get the water there. That water and energy may be making your plot richer, but it may also be making someplace else poorer.

Most of you may find, as you cover an entire block around your home, that all your neighbors have similar plants, and that greater diversity does not necessarily come with greater area. That is certainly true where I live; many of my neighbors have less affection for gardening, and break the monotony of their lot with little more than a single tree. The stream that runs past our neighborhood has been reconstructed into a neatly shaped channel with uniform banks, devoid of riffles and the life they harbor. Thus I find few new species as I take this experiment beyond the bounds of my own property. This is not true in the wild — the farther I walk through the woods, to larger tracts of land and then down to the river valley, the greater the diversity gets. Even more species appear at the marsh. Not only is there a greater diversity of species, but a greater variety of ecological systems, or "ecosystems," to provide homes for species with differing adaptations.[3]

My entire subdivision, comparable in size to the local nature park, has nowhere near the park's number of mammals, birds, reptiles, or plants. It just has more people. My survey of the subdivision roughly doubled the number of plants from my home count. This is largely because some people plant all sorts of exotic things in their yards. But the count of mammals and birds did not increase much at all. Granted, I did not go probing into people's homes to find out what kind of houseplants they harbored, or if they had pet gerbils or hamsters. At the park, however, naturalists have studied the plant and animal species for years, and the numbers compared with those in my little plot of woods become truly remarkable (fig. 1.2). As many as two hundred bird species were found in the park, and the diversity of plants was five times that of my subdivision.[4] Not counting insects, which would have been a tedious job for anybody, the subdivision pales in comparison with the nature park, which provides diverse habitats for six or seven times as many species.

What can we conclude from this thought experiment? Just our very presence, taking up space on this planet, must limit or reduce biodiversity. Keep in mind that we have not yet considered hunting and fishing. We have also ignored the vast agricultural lands that reduce once diverse lands into stretches of a single crop with relatively few ancillary life-forms. The sum of our populous existence has already led to the reduction, extirpation, and local extinction of countless life-forms, and probably will continue to do so.

Anecdotal evidence from our homes and neighborhoods points to some

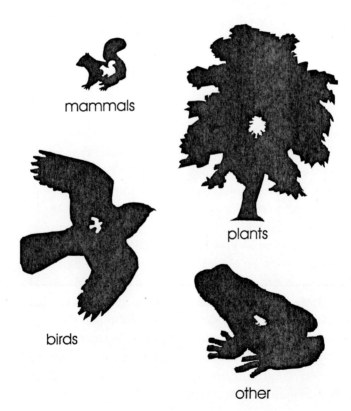

mammals

plants

birds

other

FIGURE 1.2 Relative proportions of species biodiversity in my subdivision (*white*) compared with a nearby nature park (*black*). In contrast to larger urban neighborhoods, larger areas in the wild tend to harbor a greater diversity of species.

of the basic principles of the relationship between human populations and other species. But just as every neighborhood differs, ecological systems and the human impacts on them vary as well. So we must go beyond our little experiment, and look at the broader dynamics of humans and biodiversity.

The Sparing Nature Thesis

This brings me to the first proposition of this book, which I refer to as the "sparing nature thesis." I contend that *there is a very close relationship between biodiversity loss and human population growth.* Quite simply, the more people there are, the more we push aside wild plants and animals. As our population has grown, other species have had to adapt to living in confined reserves or enclaves, lest they go extinct. But over the planet as a whole, there are fewer chances for species to survive as we continue to increase our numbers so dramatically, with a net gain of over 200,000 people every day.[5]

"Population growth seems to affect everything but is seldom held responsible for anything," wrote Princeton demographer Charles Westoff.[6] In our case this is because the relationship between human population growth and wildlife extinctions is not always direct. We are a species with an insatiable appetite for resources, and we sometimes use them irresponsibly. Thus, it could be argued, what we do and how we survive may be more important to our impact on other species than the simple effect of our great numbers.

What is it that we do? Hunting comes to mind as an example. Whereas today there are many conscientious hunters, that has not always been the case and is not representative of hunters around the world. In many cases hunting is done solely for sport or greed; even when animals are hunted for food, much of the animal is left as waste. But our wastefulness in other spheres of our behavior, such as the excessive cutting and burning of tropical forests, can lead to the demise of many species, not just animals. So in general, conservation-minded people may argue that if we did things differently, taking a kinder and gentler approach to the ecosystems that surround us, the human impact on other species would be lessened. Perhaps our great numbers could live in harmony with nature, as the Native American populations once did.

If only it were true. Near my office on the Olentangy River are ancient fortifications and burial mounds left behind by populations that lived simpler lives, and the forest withstood their existence over thousands of years to be as diverse as it is today. Moreover, the waterlogged moat protecting the front border of one of their earthworks now serves as a unique spring breeding pool for the Jefferson salamander. Sometimes human disturbance is good for certain species. But the "ecological Indian" is a myth.[7] Despite Native Americans' respect for other living things, they too had negative impacts on nature, and as we'll see, those impacts were related in many cases to substantial human population growth.

Thus we need solid evidence to establish that, over the broad sweep of time, there is a close tie between the simple count of people on the planet and the diminishing count of other species. In other words, the impact of our large population would be great *even if* we were to behave differently. If there is such a link, then it is particularly frightening, for there are now over six billion people on the planet. And our human population grows dramatically every day, as evidenced by the ever growing suburbs near my home. As a fifth-grade science student once wrote, "There is a tremendous weight pushing down on the center of the earth because so many people are stomping around there these days." The impact of our collective stomping on other species is also destined to be profound if the "sparing nature thesis" is true.

My use of the phrase "sparing nature" has two meanings. The first meaning has to do with the quote from Thomas Robert Malthus that opened this chapter. Nature has been "comparatively sparing" in the space and resources necessary to support both wildlife and a human population. Malthus was the economist who warned us in 1798 that the human population would grow more quickly than the resources necessary to sustain it. Now I know that Malthus and his ideas have been much maligned over the years by optimists who see no problem with human population growth. Thus in 1998, the two hundredth anniversary of his publication went almost without notice. The following year, the milestone number of A.D. 2000 got much more press than another milestone number: six billion, the number the human population reached sometime midway through the year. Granted, both are arbitrary numbers. But to a few of us concerned scientists, the population milestone was far more frightening than the computer glitches predicted for "Y2K." Y2K had almost no effect; the six billion figure has a lasting and growing impact. Yet few people noticed the milestone, and even fewer cared.

I am an unashamed neo-Malthusian who cares—human population growth is outstripping resources, especially as it relates to the sustainability of earth's biodiversity. As Ohio novelist and conservationist Louis Bromfield put it in 1947, "The bitter truth is that we are having our noses rubbed in Malthusian theory."[8] It is even more true today than it was then, for our population size has since doubled. Thus the second meaning of "sparing nature" is that we humans have an obligation to spare nature from the devastating effects of human overpopulation.

This brings me to a second thesis worth exploring, and one that engenders more controversy. Quite simply, if we want to conserve biodiversity on earth, *the most important conservation measure we can take is to slow or halt the growth of the human population.* Frankly, a reduction in numbers from six billion may even be desirable. Now don't get me wrong—I truly value the conservation efforts that go on today, and that contribute toward such goals. But if our human population continues to grow, then even the most ardent conservation efforts of today may be for naught.

You may have noticed that I prefaced that last thesis with an "if"—*if* we want to conserve earth's biodiversity. Is conservation really necessary? Do we need biodiversity on earth, or can we get along without it? So the black rhino goes extinct this year, and we loose a rare salamander next year. Can we tolerate that? If we can, then this investigation of the relationship between human population growth and the demise of earth's biodiversity is a mere academic exercise. It may be interesting, but of no real consequence to our survival. If only it were true.

In order to give this academic exercise more meaning, we must also establish that conservation is not only good but *necessary.* After all, if we are going to ask hungry people not to hunt the wildlife in a preserve, or ask all humans to have fewer babies, then we ought to have solid and profound reasons. Hence the third thesis to be investigated in this book: *conserving biodiversity is vital to the health of our planet, and consequently is vital to us.*

If you don't readily accept my three theses, or wonder what their scientific bases may be, read on. The evidence we'll explore in each chapter is overwhelming.

Elton's Reasons for Conservation

Many ecology text books give three main reasons for conservation, and these can be traced to the writings of ecologist Charles Elton in his 1958 book, *The Ecology of Invasions by Animals and Plants.* Elton was a strong proponent of the conservation of biodiversity, having documented its benefits for the sustainability of ecosystems, and served as an inspiration for many. The reasons found in his text are (1) religious, (2) aesthetic and intellectual, and (3) practical. Here we'll take an initial look at each, and then return to the topic in the penultimate and final chapters.

The religious reason states that animals, and for that matter all life-forms, have a "right" to exist. Whereas I have no real objection to this reason, Elton rightly noted that this "religious" or moral point of view "will seem folly to the practical Western man."[9] Many Westerners have strong feelings for the sanctity of nature, to be sure, but they unwittingly abuse nature every day. Moreover, the notion of an all-pervasive right to exist will seem even more ridiculous to anybody who hungers for the resources in a nearby nature preserve. So perhaps we should leave this reason alone, as we must also allow for diversity in people's religious perspectives.

Elton's second reason for conservation was "aesthetic and intellectual." The aesthetic side is certainly true. Most people find nature to be quite beautiful, and that is one of the reasons we plant our gardens and set out our squirrel and bird feeders. My boys love going to the woods and seeing all that the natural world has to offer. I just hope there are still plenty of woods for their children to enjoy. It comes down to this question: do we want fewer children who have more to appreciate, or more children who have less to appreciate? But, once again, there is no use in pushing the sentimentality, for many people have learned to love buildings as much as trees, see no beauty in damselflies, or find water bugs to be a nuisance. We even label vast stretches of asphalt as "scenic."

Sentimentality alone does not put food on the table for any children, thus weakening the argument for conservation. A local farmer, distraught over a conservation plan affecting her riverfront property, put it this way: "We have foxes that kill the chickens, hawks that kill the chickens and coyotes that kill the sheep. I've had enough wildlife, thank you very much!"[10] Thank goodness nobody told her that the ecologists' interests were in the rare mussels of the river's riffles — then she really would have got her dander up. But she has a point, for we compete with wildlife for our food. My response in defense of wild species: can't live with 'em, can't live without 'em. We need wildlife to ensure an ecologically healthy planet. But the sentiments of the farmer are likely to stay at odds with mine.

There is something to be said, however, for the intellectual stimulation provided by diverse species. In the words of John W. Bews, a botanist who initiated the anthropological subdiscipline of human ecology: "It is a very natural thing to enquire what bearing biological studies of any kind have on the life of man himself."[11] Not only does such research give many of us academic occupations, but we learn useful things about how the natural world works and how it applies to our own lives. This information in turn helps us grow better corn, wheat, and rice or fend off the natural predators of those crops. Yet there is much more to be learned in nature, not only in crops but in wildlife. We are only now coming to an understanding that life's "order" is much more a tangled web that can't untie itself than the simple food chain once envisioned by Elton and others. That knowledge may prove valuable. In an exam essay, one of my students wrote: "We may have won the game of evolution, but there comes a time to stop playing and learn the rules." I'm always pleased to see such wisdom at such a young age.

There are, as we are still learning, many immediate practical benefits of conserving biodiversity, hence Elton's third reason for conservation. Aside from obvious products we derive from a variety of plants, such as the rubber for tires or the maple syrup you put on your waffles, both of which are superior to their synthetic substitutes, the genetic basis of plants provides us with many tangible benefits. The usual arguments are that wild plants maintain a store of useful genetic information for our crops as well as for our medicines. Think of it this way. In a nondiverse population, such as that of a cornfield, the corn has a limited array of defenses and thus is at the mercy of the other evolving life-forms. Pests and predators, with greater variability, have ample opportunity to evolve resistance to pesticides and/or find new ways to consume the riches we have planted for ourselves. Just think of the Irish potato farmer in the 1840s to realize how a lack of diversity in crops can be devastating in the face of a new form of blight. But if we have a wide variety of plants in the wild that harbor natural resistance to pests of any sort, we can incorporate those mechanisms into our own crops, or replace the crops with other closely related plants found in the wild. Nature has done our work for us.

Likewise, a good portion of the medicines we use come from natural compounds found in wild plants. Aspirin, quinine, and hundreds more medicines are either natural chemicals derived from plants or slightly modified versions of compounds found in nature. The logical assumption is that the more plant species we allow to survive, the more likely we are to find useful medicinal compounds. The practical benefit will be longer, healthier lives.

Conservation for the benefits of "bioprospecting"—hunting for new crops and medicines—is indeed practical. But once again, a starving person who can't even afford aspirin is hardly going to be concerned about losing the medical benefits of native plants. That person will not want to suffer for the sake of curing some rich person's case of gout a continent away. One could even take an entirely cynical view that extending our lives through better diets and medicines will only exacerbate the overpopulation problem. So whereas these practical arguments for conservation hold weight for many, they are not enough for everybody. We need an even stronger reason for conserving biodiversity.

That reason is planetary health — my third thesis. The need for a sustainable planet is a reason with considerable bite. Quite simply, nature does a lot of unheralded work for us when left to its own devices. Those insects may be a nuisance, but they pollinate the plants — as do some birds and mammals. In Ohio, due to an invasive mite variant that devastated the honeybee population, we have to borrow bees for spring pollination in the orchards (more on that later). And just like the unseen microorganisms underneath the sycamore leaves at my Olentangy office, many underrated species are creating our soils, providing us with oxygen and climatic controls, decomposing our waste, and doing who knows what else.

If nature is our "mother," as is sometimes said, then surely she is like the busy soccer mom who totes her children to practice while doing much more than they can, at their age, appreciate. Nature too labors through countless thankless jobs that we take for granted with childlike innocence. Biodiversity plays an integral role in everything from climate management to erosion control. Although the systems are complex, it is time we learned the rules of the game.

Natural ecosystems are complex because of their long evolutionary history. They cannot be quickly or easily dissected with the methodical tools of science, and are not easily described with a few strokes of a scientist's pen. Short of fully understanding the benefits of biodiversity for planetary management — and we have only just begun to grasp just that — we can do yet another quick thought experiment, working from the simple to the complex. Try to imagine the fewest number of species humans would need to survive on this planet. Start with two: humans and, say, corn. That is not enough. Perhaps because I dig for a living I'm inordinately fascinated by soil, but that is the first place we need a few more species. Soil is much more than dirt — it is a complex product of living organisms, worms and insects and invisible microorganisms that ensure proper nutrient transport to the roots of our corn. We need those tiny helpers.

Corn alone does not provide us with the full range of nutrients we need for good health, so a few other crops would help. Well, unless billions of people want to spend their lives pollinating squash plants by hand, we again need a few more species to help; some insects would do the trick nicely. Too many insects, however, would be a nuisance, so some birds to keep insect populations in check would help considerably. Trees would give birds a place to land and build nests. And without trees for shade, it would be pretty darned hot on this planet. We'd also get tired of building our homes out of corn straw, despite the lack of wolves to blow our houses down.

One can go on and on with this thought experiment. Soon you'll find that to be comfortable, you will have pretty much rebuilt the world as we know it—with or without mosquitoes, depending on your preferences and level of ecological insight.[12] To imperil biodiversity is to imperil your own survival. Nature keeps us alive through what are often called "nature's services." Throughout this book we'll look at how a rich and diverse living planet maintains these services. Their practical benefits provide the strongest of Elton's reasons for conservation of other species. We can't live without 'em.

What to Do?

To accomplish our task we must first try to understand the workings of nature. This is not easy, as I have already mentioned. In many ways we simply do not know as much about biodiversity as we should like. Recent research, and general observations about our planet, certainly offer some strong incentives for conservation. But *how* to conserve is another matter. Nature has had millions of years to weave an intricate web of life, and science has just scratched the surface of these ecological complexities. Thus I am suggesting that the most assured and consequential measure would be to halt the rapid growth of the human population, and perhaps even reduce its size. After all, it is not difficult to reduce the number of pregnancies—we *know* what causes them. But because of our reproductive habits, the numbers are not going down sufficiently by themselves. We must make a conscious and conscientious effort.

Despite the tone of much of what you've just read, I am an optimist. Not in the sense of the optimists who would leave human population growth to continue, but in the sense that identifying the problems is the first step to resolving them. On the other hand, lest you become complacent that all will be well by the end of the book, let me add another caveat. Am I going to tell you that if we reduce our population, all will come right with the natural world and we will live in some sort of utopia? Heavens no. There is no one panacea to any one problem, let alone the myriad of problems involved with the biodiversity crisis. Am I going to tell you that with continued human population growth, our planet's species will be in considerably greater peril? Hell yes! And those species include one called *Homo sapiens*.

2

C H A P T E R

The Scattered Seeds

*There is grandeur in this view of life, with its several powers, having been orig-
inally breathed into a few forms or into one; and that, whilst this planet has
gone cycling on according to the fixed law of gravity, from so simple a beginning endless forms most beautiful and most wonderful have been, and are being, evolved.*

CHARLES DARWIN, *On the Origin of Species*, 1859

CHANGE is an integral part of nature. Laws of physics may stay constant,
but the only persistent law of biology is that of change. The world we live
in is considerably different from what it was about 100,000 years ago, when our
species took its modern form. Much of that change has been guided by our
own hands. But with or without *Homo sapiens*, change would have been the
norm. As one goes back in time to about 2.5 million years ago, when the genus *Homo* first arose, the world was even more different from what it is today,
with most continents as yet untouched by prehuman hands. The further one
goes back, the more vast the changes appear to be. How do we know? Fossils.

The fossil record of the past shows us that change is normal in a living
world. Life has been around for nearly 4 billion years, but the first 3 billion or
so were not all that interesting by modern standards.[1] Our earliest fossils, dating to 3.5 billion years ago, show us that initially there existed little more than
simple bacteria-like creatures on the planet, and they diversified over the next
couple of billion years. By about 2 billion years ago, these single-cell creatures
evolved into larger, more complex cells. It was slow going, but simple cells are
not so simple, which is why more highly evolved bacteria still dominate the
living world today in many ways. From the early pioneering cells there was
still a long road of evolution before the basic patterns of biodiversity as we
know it could become established.

Eventually the stage was set, and the last half billion years or so of earth's long history became considerably more interesting (to us, anyway). About 540 million years ago, life reached a threshold of complexity, and the diversity of life blossomed. At this time, during the so-called Cambrian explosion, the creatures of the ancient seas became a bit more recognizable as forms akin to us. It may be hard for some to look at a trilobite (fig. 2.1) and see a pattern familiar in our own bodies, but these sea creatures were segmented and bilaterally symmetrical like us, and had the basic divisions in their bodies of a digestive system, central nervous system, and so on. The stage was set.

Subsequent appearances of reptiles, amphibians, and dinosaurs moved the changing world closer to the diversity of today. Life of the distant past was diverse as well, though the dinosaurs' world of 200 million years ago would hardly have been recognizable to us today. That is perhaps what is so astounding about the fossil record — it shows us that most of the life-forms that ever existed are now extinct. Within my own field of evolutionary anthropology, there are debates about how many species along the human lineage became extinct. Certainly some species went extinct. There is no doubt that the "robust australopithecines," cousins of ours with large teeth and faces, met

FIGURE 2.1 The trilobite (*left*) was an early sea creature that showed some basic features of modern animal form and shape. Now extinct, it was part of a vast community of plants and animals (*right*) that lived in the sea during the Cambrian period, setting the patterns of biodiversity for further evolution over the next half-billion years. (*Illustrations by Michael Masters*)

their demise in the evolutionary wringer. More familiar to you may be the Neandertals, now gone by one or another form of extinction.

No matter what happened in terms of species extinctions, certainly genetic variants were lost, including the genes that gave the Neandertals their peculiar features, so similar yet distinct from ours. That is the very nature of evolution through natural selection: it is a pruning process. We have inherited a diverse set of genetic leftovers that survived the rigors of evolution. What is of concern to the human population now is the nature of the process and its sheer magnitude.

Far from being the simple death of a species, extinction takes many forms and has many potential causes. Sometimes there are mass waves of extinction that wipe out the majority of living plant and animal species. There have been five major "mass extinctions" in the past, including the most recent one, which killed off most of the dinosaurs (fig. 2.2).[2] But one need not have an asteroid hit the earth, as apparently happened near the end of the dinosaurs' reign, for there to be extinction events large or small. Other extinctions have been more subtle, and a species quietly goes extinct as soon as it can no longer keep up with the changing world. To be sure, new species arise through the evolutionary process and carry on the business of generating life. But the extinction of a species is always forever. They can't come back.

Despite the permanence of species extinctions, one kind of extinction carries more hope than others. This is called a "transitional extinction." As a population evolves into a new form, the old characteristics fade away, and we recognize the successors as a new species. The lineage did not die out, just some of the features of the former species. So whereas the fossil species recognized as *Homo erectus* may be extinct today, it seems to have produced a lineage that led to all of us (which may or may not have included the Neanderthals). The extinction of the species is just as permanent—*Homo erectus* cannot come back—but there was a transition to something new.

No matter the form, evolution and extinction are parts of natural change and, in one way or another, affect the diversity of life. Thus, in order to understand the dynamics of earth's biodiversity today, we must first understand evolution.

Darwin's World and Welcome to It

Evolution is a marvelous phenomenon that is responsible for the diversity of life as we know it. Through studies of the fossil record and comparative

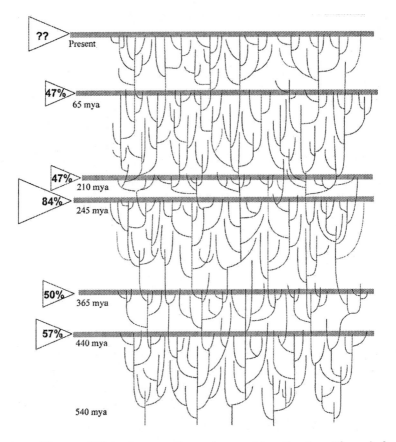

FIGURE 2.2 The tree of life has been continuously pruned by extinctions. The end of each line, or branch on the vertical "trees," represents the extinction of a lineage. Five times since 540 million years ago (mya), when the basic patterns of life's biodiversity were established, there have been concentrations of extinctions (*shaded horizontal lines*) when greater percentages of living forms were extinguished. Today we may be going through a sixth mass extinction.

research on the genetics, morphology, and physiology of living species, the scientific community has a good grasp of the processes and patterns of evolution. We've come a long way since Darwin's time. Indeed, the importance of evolutionary theory for understanding life has made evolution the cornerstone of modern biology. It is not my intention here to describe evolutionary processes in detail. Rather we need to elucidate how Darwinian evolution leads to biodiversity, and how extinction is part of the process.

Although evolutionary theory is fundamental to modern biology, popular misconceptions about Darwin and evolution abound. Many people have not had the educational opportunities or time to think about evolution. It is no wonder that such people do not understand the magnitude and significance of extinctions going on today, for an understanding of evolution is essential to interpret our current crisis.

The question I get most often from those who have been relatively unexposed to Darwinian concepts is this: "If we evolved from the apes, why are there still chimpanzees and gorillas around?" The implication is not necessarily the naïve view that we evolved from modern apes; most people know better than that. Rather the thought going through their heads is that once we split off from the apes, as an obviously superior species, our closest relatives should have been driven to extinction in the Darwinian struggle. I always jump right back and note that we *are* driving modern apes to extinction, but that is rarely the answer they seek. The underlying misconception is that the evolution of one species always leads to the extinction of another — that it is a totally linear process. It is more complicated than that, and at the heart of the answer is the key to the origins and maintenance of biodiversity.

It is worth briefly following up on the chimp and human saga. Humans and the African apes evolved from a common ancestor, around five or six million years ago. Whereas that common ancestor is now extinct, it is quite possible that it lived on for some time after it generated a somewhat different daughter population. One population of that species accrued changes that put it on a unique path toward walking upright and growing a larger and more complex brain. Another population of the ancestral species, probably located on a different part of the African continent, took another trajectory toward knuckle walking and other behaviors and shapes.[3] These two adaptations were not mutually exclusive, and both lineages continued to evolve. In some ways, gorillas may even be "more evolved" than we are, in the sense that they seem to have acquired a greater number of genetic (and perhaps morphological) changes than humans in the time since we diverged from a common ancestor.[4]

Whatever happened in the details of the process, the evidence shows that the group of so-called higher primates differentiated from one species to at least three. Further down the line, even more species arose and went extinct. We see remnants in the fossil record of numerous species that looked vaguely like us. The species diversity of our family tree increased. Those that could navigate the course of an ever changing world continued on to today; others followed the norm known as extinction.

If it were not for this splitting of lineages, or "evolutionary divergence," the world would be a very mundane and monospecific place. From amoeba to human, only one species would be alive at a time. Evolution of creatures such as mammals never would have been possible without divergence, for they would have had nothing to eat but their own kind. As it is, a vast array of species evolved into a complex web of those who eat and those who get eaten. The plants and animals continued to evolve into ever more intricate relationships in which one species may depend on many others. It is this diversity that keeps us alive — *all* of us. And it is precisely the reason that there could be no world with just humans and corn.

Every species alive today is the finely tuned product of millions of years of evolution. One is sometimes tempted to see nature as being "red in tooth and claw," with clever or nasty organisms outcompeting others to the death.[5] Natural selection, or the "survival of the fittest," is sometimes ugly. Whereas such scenarios do play out as part of the process, the Darwinian source of biodiversity is more a simple "birds and bees" explanation of life writ large — those who could live to produce viable offspring continued on. Success and adaptability bred more success, thus being naturally "selected" for continuing generations. The complex ecological systems we now live in are products of billions of years of plants and animals successfully adapting with each other and to each other. There is indeed grandeur in this view of life.

Far from being ugly, evolution through natural selection has built the beautiful interrelationships between bees and pollinated flowers, worms and the roots of vegetation, or gorillas and the rich vegetation of the hillsides where they live. The latter example is particularly rewarding, albeit unusual. Gorillas, like humans, are often quite playful. Mountain gorillas roll down hillsides with each other, with antics reminiscent of our own children. In the process, they roll over woody saplings of young trees, thereby preventing a heavy canopy of trees from growing. This allows the herbs they consume to grow in the abundance of sunshine; indeed, the herbs depend on the gorillas.[6] Moreover, gorillas are pretty good conservationists in some respects — they carefully crop some of the vegetation they eat, leaving enough of the plant to encourage new growth.

The more life-forms there are living together in an ecosystem, the more complex the interrelationships become. One of innumerable examples: a successful species of tree may depend on insects for pollination, monkeys to spread its seeds, birds to eat potentially damaging insects, vegetation on the ground to keep the soil from eroding, and microbes to nourish its roots. Not only has there been evolution within species, but there has been coevolution among species — a process of intertwining species' adaptations.[7] These complex systems took millions of years to evolve, and the resulting web of life is

not easily unwoven. The extinction of one species may have dire consequences for its coevolutionary partners. Thus, in today's highly evolved world,
it takes biodiversity to sustain biodiversity.

In looking at the intricate web that has evolved, and what humans are
doing to it today, a basic question emerges as a starting point for understanding: just how much biodiversity *has* evolved? Finding an answer to this simple
question is more complex than one might think. Yet, to phrase the answer as
succinctly as possible: we don't quite know how much biodiversity has
evolved, other than that there is a lot of it. How much is still evolving and how
much is going extinct? For the time being, we have to rely on estimates. But
the current picture, though not precise, is coming into focus from a wide
variety of research endeavors. Let me put you into the picture.

Counting and Recounting Species Biodiversity

My favorite place in the world, outside of Ohio, is the Makapansgat Valley of
South Africa (fig. 2.3). I spend my time there because of the fossil riches —

FIGURE 2.3 The Makapansgat Valley of South Africa holds a wealth of biodiversity today,
along with fossils revealing an even richer variety of life over three million years ago.

bones in ancient cave fills, sporadic samples of life's changes over the past three or four million years. But the valley is of great interest to more than just paleontologists. It is an area rich with biodiversity for scholars, soil for farmers, and water for plants and animals and humans. Deep in the valley, there has been very little human encroachment—yet.

The valley could be considered a "catchment" area in many senses. It is, for me, a fossil catchment, hence its value for excavations. More relevant to the people living there today, the Makapansgat Valley is a water catchment where the rains run from the hills and the upper pools down to the streams at the base, and fill the water tables below. Where there is water there is life, and in this case plenty of it. So the Makapansgat Valley is also a biodiversity catchment.

The life in this valley is astounding. I spend much of my time watching the primates: the baboons and vervet monkeys during the day, and the bush babies (galagos) at night. These animals diverged and subsequently adapted to share different parts of the Makapansgat catchment. Other primates, our distant ancestors whose fossil remains we find, could have found an enormous amount of plant food sources ranging from seeds and tubers to fruits and berries.[8] However, these prehuman ancestors had more primate competitors for food in the valley, for many of Makapansgat's ancient primates have since gone extinct. Nevertheless, the enclave is still rich with biodiversity—for now.

I was thoroughly intrigued during one of my field seasons in the valley when we were joined by two entomologists from a local university. Whereas I may have difficulty finding fossil hominids from the ancient solidified cave in-fills, the entomologists never have any trouble finding the source of their scientific passions: insects. One can always find bugs. In just one weekend, they captured four or five species of insects that had been unknown to science, and returned to their labs to describe the unique traits of each kind. On subsequent visits, they assured me, there would be no problem in finding more new species. These are not new in the sense of being recently evolved, but new in terms of being named and described for the scientific community.

Today, across the world, biologists are still finding new species, one at a time. They are not just teasing out rare or elusive insects. Even though mammals are probably the best-documented group, two new species of monkey were discovered in the Amazon jungle at the close of the twentieth century.[9] Despite hundreds of years of biological research on every part of the globe, there are more species to be discovered of all types. That is, *if* we can find them before they go extinct.

It seems peculiar, sometimes, that we can't find and identify new species more quickly. In the early nineteenth century, Darwin accompanied mappers on the famous voyage of the H.M.S *Beagle*. As they slowly mapped coastlines of continents such as South America, Darwin went ashore and began the painstaking task of collecting and identifying the abundance of species. Darwin had plenty of time, for it took a course of years to make maps of even limited accuracy. Today a satellite can make a detailed map in a matter of minutes, while the biologists still have to traipse through dense jungles to document life's riches.

True, biologists can also use satellite imagery to estimate the amount of biodiversity in an area. Areas rich in biodiversity tend to be more productive in assimilating energy from the sun, and that energy is measurable from satellites in space. Thus we can see greater productivity in an Amazonian jungle than in a subtropical grassland, and can quite assuredly state that the jungle harbors greater biodiversity. This makes sense from land observations as well. Likewise, a North American woodland will show up on a satellite image as having greater productivity than large stretches of farmland. Despite our vaunted agricultural productivity, it is nowhere near that of a natural woodland. Such observations from space may give us estimates of the species numbers of particular areas, but not of *which* species live there. We still must find them by foot, one species at a time.

It is estimated that there are well over a million known species of plants and animals. Again it seems peculiar that the number of *known* species should have to be estimated. But there is no central catalogue.[10] Once a species is identified, it is published in one of many international academic journals. Usually a single representative, the so-called type specimen, is then housed in a university or museum for future reference. But it is not always possible to cross-check these species across every academic institution, so a single species might be named twice and have "type specimens" at two different places. Keeping that in mind, and after scouring the vast academic literature, scholars have assembled the figure of 1.4 to 1.6 million "known" species.[11]

The unknown is what scientists strive to explain, and the total number of species is not known. But given a reasonable estimate of the number of species, known and unknown, it would take about seven hundred years to identify all of them at the current rate of discovery and description. What is that "reasonable" estimate? About 12.5 million species on earth. In other words, there may be nearly eight times as many unidentified species as there are species recorded by scientists. Or there could be even more. It is worth looking at how science has come up with such figures, so that we can understand the vastness of biodiversity.

Many estimates of total species diversity have been made, using a variety of techniques. All are based on one simple principle: extrapolate from the known to the unknown. Here I'll give three examples. Peter Raven is a biodiversity expert from the Missouri Botanical Gardens in St. Louis. He observed that in temperate regions such as North America there are fewer species than in tropical forest areas. Certain groups of animals such as mammals and birds are fairly well documented in all areas (despite the few monkey species that get discovered on occasion), and the tropical regions have roughly twice as many species. If that holds true for other kinds of organisms, then one just has to take a map or a satellite image and start doing some calculations to extrapolate from the known species to the unknown. Raven came up with a figure of three million plant and animal species on earth, one of the more conservative estimates.[12]

Nigel Stork of Australia and Kevin Gaston of the United Kingdom arrived at a different estimate using a similar form of extrapolation, but working with intimate knowledge of butterflies and other insects in Britain. As butterflies are fairly well studied worldwide, one can get a pretty close estimate of the total number of butterfly species. In Britain there are 67 species of butterfly and about 22,000 species of other insects. Now we can use that ratio of butterflies to insects to do some more math. Assuming that the same ratio holds worldwide, and knowing that the world has 15,000 to 20,000 butterfly species, that would mean a sum of 4.9 to 6.6 million insect species alone.[13] That number dwarfs Raven's total species estimate of 3 million.

Other estimates of the number of insect species have been much greater. A famous, or perhaps infamous, study by Terry Erwin of the Smithsonian Institution put forth a much larger and more astounding figure of 30 million species of insects (technically arthropods) in tropical forests alone! How did he come up with that? From a mere nineteen trees in a tropical forest, he counted 1,200 species, most of which were beetles. He didn't just wait for them to fly or crawl by, as I did for my amateur experiment in the woods of Ohio. He used a "knockdown" insecticide in the trees and identified the fallen bugs. Many of these were specifically adapted to certain kinds of trees, whereas others were adapted to larger areas. Taking this all into consideration, with some complicated math, he got his global figure of 30 million.[14] The study has been heavily scrutinized in the global scientific literature, and the consensus is that Erwin's figure is an overestimate. Fair enough, for he had to make a lot of assumptions in order to do the math. On the other hand, there were probably still more insects in the ground that he did not count. If nothing else, his study shows the difficulty of making an accurate estimate of biodiversity—and that a whole lot of insects are out there.

Yet we don't know exactly how diverse plants and animals are. It's ironic that we should spend billions of dollars to investigate the possibility of life on Mars when there is so much left to learn about life on earth. Like many people, I was enthralled watching Sojourner, the little rover, probe soil and rocks on the Martian surface. But the earth has a more important set of questions: how many species are there on *this* planet, and how many are going extinct? We are left with contrived estimates.

Most scholars would agree on an estimate that our earth has between 5 million and 15 million species (not counting bacteria and viruses). As a single number, the estimate usually converges on about 12.5 million. Of course the truth cannot be obtained by scientific consensus—the real figure has yet to be determined. But the estimated figure can give us a number to work with, and give us an appreciation of the diverse products of evolution.

One reason we want to know the number of species is that we want to determine how many are going extinct. Is the proportion of extinctions large or small? Is it normal or abnormal? As it turns out, documenting extinction is as difficult as—if not more difficult than—counting living species. All we know now is that there have been a lot of species extinctions in historical times. Again, biologists must find them—or rather, find out they are missing—one at a time.

The End of the Road

Perhaps a visit to the rain forest of western Ghana to view Miss Waldron's red colobus in the wild was not high on your priority list, but it appears to be too late now anyway.[15] This type of monkey was abundant in the 1950s, two decades after Willoughby Lowe found the distinctive creature that was to bear the name of his companion, Miss F. Waldron. In extensive surveys conducted from 1993 to 1999, a team of primatologists failed to locate the monkey. It is not as if they were looking for a shy, retiring animal that could easily hide— Miss Waldron's red colobus is, or was, a large, conspicuous monkey with distinctive vocalizations (hence its appeal to the local hunters).

According to my Ohio State University colleague Scott McGraw, the silence he witnessed in the forests of western Ghana and the eastern Ivory Coast was "eerie." The remaining tracts of natural lands searched by McGraw and his colleagues looked like normal, healthy forests, but were disturbingly quiet. "A healthy rain forest is *loud*," he told me. "We managed to locate one group of Diana monkeys, which is a species notorious for its raucous behavior. In the Tai forest of the western Ivory Coast, this species is easily found because group members are constantly chattering away with each other using squeaks, whistles, and barks. The group we found in the east was found by accident — we never heard them. We were amazed to find that this group was foraging while making virtually no noise at all. It seems they learned that in order not to get shot, they had to keep their mouths shut. Unfortunately, it looks like Miss Waldron's red colobus was unable to make this adjustment. Simply being a large, red, noisy monkey made it easy to find and kill. And now there may be no more."

The exasperated team declared the monkey likely to be extinct in 1999 — the first known extinction of a primate in the twentieth century.

The twenty-first century is likely to see more extinctions of Miss Waldron's cousins, for their homes in western Ghana and the eastern Ivory Coast are under siege. Hunting appears to have been the main reason for the extinction, as the local human inhabitants valued Miss Waldron's red colobus for "bush meat." This despite thirty years of warnings from scientists who had visited the area. The outlook for other primates does not appear to be getting better. Indeed, the team of scientists was offered a smoked specimen of a related monkey for the bargain price of three dollars. The problem promises to grow. To quote the team: "Hunting probably has been exacerbated by the improved access to forests resulting from logging activities and by the growth of the human populations in the area formerly inhabited by the red colobus."[16] The extinction was in part a product of what humans do, and in part a product of how many of us are doing it. We moved in, something else was forced out — forever.

One extinction does not a crisis make. But it does give a name and a face to the growing list of extinctions. And, again, we have to rely on estimates in order to measure the extent of the extinction problem. Worldwide, there are only about one thousand species that are *known* to have gone extinct in the past five hundred years. Only about eighty eight of those extinctions were mammalian species.[17] That may sound sad, but not terribly alarming. On the other hand, the words of Michael Soulé are rather troubling: "As far as I know, no biologist has documented the extinction of a continental species of plant or animal caused solely by nonhuman agencies such as competition, disease, or environmental perturbations in situations unaffected by man."[18]

Despite the seemingly small numbers, it must be emphasized that we are dealing with known, documented extinctions. As was the case with Miss Waldron's red colobus, it is not easy to "know" that something is extinct. And, as with the count of living species, the true count of extinct species is likely to be much greater than the known count.

Whereas species extinctions are difficult to count, we should not jump to the conclusion that the scientific community has not garnered enough evidence for true concern. Labeling ecologists as "alarmists," retired environmental consultant Rowan Martin states: "Those losses of species that have been clearly documented do not appear to have dire implications either for ecosystem functioning or for human survival."[19] Appearances can be deceiving, as we'll see. Paul and Anne Ehrlich, who have spent part of their careers answering the critics of environmental science, put it this way: "Biologists don't need to know how many species there are, how they are related to one another, or how many disappear annually to recognize that Earth's biota is entering a gigantic spasm of extinction. All they need to know is that high rates of habitat destruction and alteration are occurring everywhere and that most species have quite limited distributions and are highly habitat specific. The conclusion is obvious."[20]

Throughout the following three chapters, we'll approach an honest accounting of species extinction, and get closer to the "terribly alarming" stage of knowledge. But enumerating species is just one way of looking at biodiversity. If we also look at biodiversity above and below the species level, we can better understand the making of a crisis.

Hello Dolly

A species is one kind of category contrived by scientists to understand the diversity of life. It is sometimes difficult to define "species" boundaries. Sure we recognize *Homo sapiens* as one species, and *Pan troglodytes*, the common chimpanzee (which is becoming less common every day), as another. But the baboons I watch at Makapansgat are somewhat different from those of the South African coast, and even more distinct from their close relatives in East Africa. Do these baboons represent one species, or two, or three? It depends largely on opinion and one's definition of species. The importance for our discussion here is that there is always diversity *within* a species. Thus a true appreciation of earth's biodiversity must reach to a deeper level, down to intraspecific diversity of genes and other characteristics. Each human individual is different from another, as is each baboon or each sycamore tree.[21] Species themselves are a rich source of biodiversity. Among humans the pruning of genetic variability would be called "genocide"—there is no reason that the term cannot be applied to other species.

In order to understand the importance of diversity within a species, it is worth looking at the extreme case—a situation with almost no genetic diversity. This is the case with many of the world's crops. Out in our fields are "monocultures" of corn or wheat; a monoculture is an extensive area in which one variety of crop is grown. It has very little genetic diversity, leaving every plant in the field susceptible to the same hardships, be they disease or drought. Monocultures invite the ugly side of natural selection: a swift and total demise. But with greater crop diversity, some plants are more likely than others to survive particular threats of the natural world. Genetic diversity thus provides the beauty of life in Darwin's world, and the security for life's sustainability.

Despite our understanding of the value of intraspecific biodiversity, farm animals seem to be on their way toward becoming the mammalian equivalent of corn monocultures. As it stands now, farmers have bred a lot of the diversity out of cattle and sheep. That is why we cannot expect supercows to appear with even greater milk-producing capabilities—they've already reached their genetic limit. How can we limit the variability further? Cloning.

In 1997 the world was somewhat astonished, concerned, and a bit amused when Scottish veterinary researchers of the Roslin Institute announced the arrival of a cloned sheep named Dolly. This ewe, a fatherless facsimile of her mother, was the vanguard of something that has now become commonplace. Already, the cloning of animals has become less astonishing to the public. On this subject I must agree, in part, with the wisdom (and wit) of essayist Wendell Berry, who wrote: "Cloning, besides being a new method of sheep-stealing, is only a pathetic attempt to make sheep predictable. But this is an affront to reality. As any shepherd would know, the scientist who thinks he has made sheep predictable has only made himself eligible to be outsmarted."[22] True, but as a scientist, I can make *one* assured prediction: if we were to move toward producing monocultures of farm animals, the lack of diversity would imperil the flocks and herds. Good-bye Dolly.[23]

The same is true in the wild. A lack of genetic diversity is a liability, for it makes a species less sustainable—less likely to survive, say, a new kind of virus.

And therein lies a significant part of the biodiversity crisis. In most cases, the larger the population of a species, the more diverse it is. Six billion humans are amazingly diverse, as you can tell from a quick look at your neighbors or even your own family. The 200,000 or so remaining "common" chimpanzees are considerably less diverse, and the 600-odd mountain gorillas in the wild are frighteningly less diverse still. Yet their very future, their potential to biologically adapt to the ever changing world around them, depends on such diversity. The fewer their numbers, the fewer chances they have to survive. This is why the remaining monkey species of West Africa are heading the same route as Miss Waldron's red colobus.

One need not go to Ghana to find imperiled species. The extinction of genetic biodiversity is a global phenomenon, and happens in your backyard — and I mean that literally, as I outlined in the first chapter. Anywhere habitats are reduced or destroyed by human expansion, the numbers go down in wild populations, and genetic biodiversity suffers. Thus conservation conscientiousness must extend beyond saving a species, to saving the *sustainability* of a species and their many populations.

I'm not done with biodiversity yet. There is still at least one more way to look at the evolution of biodiversity — and its demise.

Bogged Down in Biodiversity

About ten thousand years ago — just yesterday, in geological time — the part of Ohio where I was to be born was partially covered with ice and snow. Year after year of ice age snow had built into massive glaciers. The weight of the accumulation pushed the immense glaciers southward, leveling the land and gouging out basins large and small. The edge of the glacier reached midway through the area of my home county before a global warming trend caused it to melt and recede. As it melted, the waters carved today's river valleys and filled the basins, the larger ones becoming the Great Lakes of North America. But it is a smaller pocket left behind in the glaciated portion of my home county that concerns us here.

One small kettle hole lake was left with sandy, acidic soil, sheltered from the glaring sun that warmed and dried the land of Ohio. An odd assortment of plants, particularly well adapted to the Spartan but wet conditions, found their way to the lake edge, grew, and flourished. A bog was born.

Ohio used to have many such bogs, but as farms spread across the land to exploit the soil that was the heritage of the glaciation, one by one the bogs became subsumed in fields of onions and potatoes. Few remain today, only to be found by those who look carefully.

My eldest brother Steve, a botanist and naturalist by trade, was one who looked. It took him years of searching, sleuthing among historical records, and probing the memories of old-timers who knew the land well. Eventually he found a bog. It was a small but rich reward, for it was filled with rare carnivorous plants and delicious cranberries — a botanist's treasure.

The bog has value to more than just a botanist. It is a unique habitat harboring a wealth of life, providing sustenance not only for its peculiar plants but for the birds and animals that visit it. A bog is everybody's treasure, for it is part of the world's heritage of higher biodiversity: ecosystem biodiversity. To make a long story short, having failed to get the state of Ohio or various conservation organizations involved in preserving the bog, my family bought the patch of land where the bog had been conducting its unique ecological functions for ten millennia.

Why save the bog ecosystem rather than transport its rare species to a safe haven? One cannot just take a rare species from a bog, plant it somewhere else, and expect it to grow — it is not a simple matter of add water and stir. A bog plant requires its own particular conditions, in this case formed by a long-term process. But this is true of all plants and animals, excepting modern humans. Species require particular habitats with the necessary ingredients to support their ways of life. Some habitats are large, some are small. *Homo sapiens* found ways to extend their habitats with cultural means, by harnessing energy and creating artificial shelters. But the vast majority of life on earth does not have the luxury of culture. The places that nonhuman species live are in effect dictated by nature — by ecological systems that have evolved over thousands and millions of years. In order to sustain species biodiversity, one must make sure that ecosystem diversity is also preserved.

Science is well aware that all ecosystems were not created equal, at least in terms of species biodiversity. We find biodiversity virtually wherever we look for it, but in different proportions. Warmer, wetter places harbor the greatest number of species. That is why many efforts of conservation are aimed at tropical rain forests, for the tropics have the requirements for high biodiversity: energy and water. Generally, as one moves away from the equator, one finds less biodiversity. The northern tundra can harbor fewer species — there is less solar energy to fuel the plants upon which the web of life is built.

Between the two extremes of equator and the poles, however, there are many variants on the theme. In Africa, for example, traveling north from the equator quickly puts you in desert, with low biodiversity. The same is true going south, at least on the western portion of the continent. Farther south, in my former home of South Africa, one again encounters high biodiversity close to the Cape coast. Indeed, the "fynbos" of the Cape Floristic region near Cape Town is among the world's most diversified plant habitats. The difference: water. There might not be the high energy of the equatorial regions, but high biodiversity needs both energy *and* water.

Traversing Africa in any direction, one thus encounters many different ecological systems determined by water and energy as well as parameters such as substrate, altitude, and so on. When Westerners envision Africa, pictures of zebra and giraffe running across the savanna grasslands are usually first to come to mind. Oddly enough, the tall-grass savanna is a habitat largely created by humans, for the grasslands must be set afire in order to be maintained. Much like the gorillas who roll down hills, break saplings, and thus ensure the growth of their favorite herbs, humans set fires to destroy the competitors of the tall grasses and maintain good grazing for their cattle herds.

African savannas are only one type of ecosystem. Just driving from the "high veld" city of Johannesburg to my fossil site in the forest of the Makapansgat Valley, I encounter not only savannas but wetlands and mountain habitats, each with its own set of typical and rare species. Africa, like North America, also has a series of great lakes (created by the cracking of the continent along the Rift Valley rather than by glaciers). Elsewhere are the remarkably diverse swamps of the Okavango Delta in Botswana. It is a continent rich in ecosystem biodiversity. Most continents are. That is, in part, why larger areas tend to have greater species biodiversity as well.[24] Each species can find its own habitat.

The difficulty is that many of these ecosystems are attractive not only to particular plants or birds or mammals but also to the mammal known as *Homo sapiens*. In the case of the African savanna, humans have become a part of the ecosystem. It works out reasonably well, for pastoralists by their nature remain in fairly small numbers and easily share the vast savannas with other creatures. In other cases, the ecosystems have to survive beside growing numbers of farms, logging camps, roads, or tourist hotels. The growing numbers of humans make it that much more difficult for a species to maintain its place in its habitat.

North America has richly diverse ecosystems as well. Among those which are the most diverse in terms of species are wetlands, much like our bog in

Ohio. Not only rare plants but migratory birds need these wetlands for at least part of the year. A host of reptiles and insects also require wetland habitats for their survival. Wetlands in North America, however, are at a distinct disadvantage. They happen to exist where people want to be. Here in Ohio, the bogs were not the only casualties. Northwest Ohio was once largely covered by the "Black Swamp," which was about the size of the state of Connecticut. But it was in our way and had all those pesky insects, so under the U.S. government's 1849 Swamp Lands Act, the swamp was drained and converted to agricultural and industrial land. It was gone by 1885. With it went a huge chunk of our ecosystem biodiversity.

One can drive down the East Coast of the United States and see hundreds of beautiful golf courses where wetlands used to be. Wetlands insects such as mosquitoes, and reptiles such as alligators, are seen to be a bit of a nuisance for a leisurely eighteen holes, so every effort was made to alter the lay of the land for greater human comfort. Somewhat belatedly the U.S. people and their government saw the growing impact of wetlands destruction and decided to do something about it. Now the law is just the opposite of the 1849 act.

But without a true understanding of biodiversity and its evolutionary origins, the lawmakers can leave huge loopholes for developers. One such law stated that a wetland could be drained or altered as long as another one was created. Savvy developers went out and bought up inland farms, flooded them, and voilà, a wetland! As if "add water and stir" would work. Of course none of the species of the natural, highly evolved wetlands were there. Nor will most of them ever inhabit the artificial wetlands. One cannot simply transplant an ecosystem. Nature may have "scattered the seeds of life abroad with the most profuse and liberal hand," but those seeds must find the right soil.

I am not trying to critique human development (I don't think I am, anyway). Indeed, I like to eat food grown on farms in northwestern Ohio, and occasionally enjoy playing a bit of golf along the coast. I just want to make one point: there is a relationship between human expansion and biodiversity loss. In this case we are losing part of our ecosystem biodiversity, and with it goes species biodiversity and the genetic biodiversity of the surviving species.

It seems that the ecosystems that harbor the greatest species diversity also tend to attract the most people. We are, after all, an animal that shares habitat preferences with many other animals. So the focus of conservation today is often on the wetlands or the tropical forests. It is my opinion that less diverse ecosystems, though not equal in numbers of species, are equally valuable. Every ecosystem is the fine-tuned product of the evolutionary process.

And the earth needs ecosystem biodiversity in order to maintain the business of supporting life (including human life). An arctic wilderness or a desert may not harbor a lot of different species, but they too perform a function on this planet as habitats for a special few.

Return to Darwin's World

Biodiversity at all levels — genetic, species, ecosystem, and everything in between — is a product of evolution and extinction. The process is normally slow and hard to see. But those who take the time to observe, like the team of primatologists in Ghana, can see distinct change.

It is not just extinction, as evinced by Miss Waldron's red colobus, but evolution that continues as well. One example from the state of Washington demonstrates a relatively fast and visible evolutionary event among sockeye salmon. These salmon were first introduced to Lake Washington in 1937, and their populations quickly grew. By 1992, it appeared that two distinct species had evolved from the original stock, one living in the lake near the beach, the other preferring the river tributary. These species were replete with appropriate size and shape differences, adaptive for their respective habitats.[25] All in a maximum time period of just fifty-six years.

Darwin's world is something to marvel at, but it is also something to be concerned about. Evolution is not normally as quick as it was for the salmon. And the pace of extinctions is increasing. It has been for thousands of years. The reason is the ugly side of evolution.

Charles Darwin provided an apt analogy in his introductory text of 1859 on the subject. Said he: "The face of Nature may be compared to a yielding surface, with ten thousand sharp wedges packed close together and driven inwards by incessant blows, sometimes one wedge being struck, and then another with greater force."[26] As each wedge is driven in, another falls off (fig. 2.4). Normally the wedges are thought of in terms of variants within a species, with natural selection delivering the blows to ensure placement of the "fittest" wedge. But the process can be thought of on a different level, with the wedges being more successful species taking the place of others in an ever changing world of limited capacity.

Whether it be mammals outcompeting dinosaurs, or successful early members of the genus *Homo* wedging out some of their remaining primate cousins, species compete for resources. Ultimately, they compete for lifeways

FIGURE 2.4 In Darwin's wedge analogy, as one species is driven into an ecosystem, others may be forced aside and into extinction.

within ecosystems of restricted boundaries and on a planet of limited resources. That is why so many species have fallen off the edge of nature's "yielding surface" and littered the fossil record with extinct creatures.

On the other hand, nature's "surface" has yielded to the evolution and coevolution of a diverse set of organisms. Each evolving organism need not totally wedge out another, and may indeed provide a coevolutionary opportunity for another wedge. The successful inhabitants coexist by sharing resources—by becoming a part of the ecosystem's intricate web. But when one species takes up more than its share, it may disrupt the system and lead to widespread extinction.

Biological evolution and extinction are the norms on planet Earth. Species have always gone extinct, and new ones have always evolved. It will be my contention throughout the rest of this book that these norms are now out of kilter because of one particularly successful and populous species, *Homo sapiens*. Humans appear to be the ugly reason that the current pace of extinction is far greater than the pace of evolutionary origins. In order to provide evidence, I'll trace the growth of the human wedge from its inauspicious beginnings to its enormous impact of today. We'll find that the diversity of ecosystems is in peril from the lack of species biodiversity, and species are jeopardized by the lack of genetic diversity. And, with any luck, we'll find both the cause of, and the cure for, the symptoms displayed by an unhealthy planet. I'll give you a hint—this time the cause is not an asteroid, and the cure is within our reach.

3

C H A P T E R

The Human Wedge

THE warm and wet tropics are a cauldron of biodiversity, so it is little surprise that the ancestry of humans can be traced to the tropical lands of Africa. The vast majority of our primate cousins, such as monkeys and apes, now live in the tropics. Such was the case sometime between six and three million years ago, when one primate lineage now known as the genus *Australopithecus* made its evolutionary debut and traversed parts of the African continent (fig. 3.1). Given the global spread and domination that characterizes the eventual descendants of *Australopithecus*, one might expect to find evidence of a fairly impressive animal, poised to rule the world. But our ancestral line had a most inauspicious beginning with a peculiar and meek creature.

What was *Australopithecus*? It was a somewhat apelike, mostly vegetarian creature with a few evolutionary novelties. For one thing, as best we can tell from the fossil bones, *Australopithecus* stood upright and walked on two legs. Indeed, the 3.8-million-year-old fossil site of Laetoli in Tanzania has remarkably preserved the footprints of three *Australopithecus* individuals who traversed some recently lain volcanic ash. But they did not tower over their contemporary animals, for evidence from the famous Lucy skeleton of Ethiopia (3.2 million years old) shows that they were about three and a half feet tall.[1]

Robert Louis Stevenson wrote: "Despite our wisdom and sensible talking, we on our feet must go plodding and walking."[2] The plodding and walking is our heritage from *Australopithecus*; the wisdom and sensible talking came only later in our evolutionary history. Stevenson's apt phraseology becomes apparent if you have ever tried to run from a charging carnivore. Be it a leopard or your neighbor's dog, you will be aware that two plodding legs don't move you very fast. Moreover, *Australopithecus* did not have large threatening canine teeth, such as those of modern baboons, to ward off predators through facial threat gestures. It is a wonder that our ancestors survived at all. Perhaps they were aided by a modicum of brain expansion, but this went little

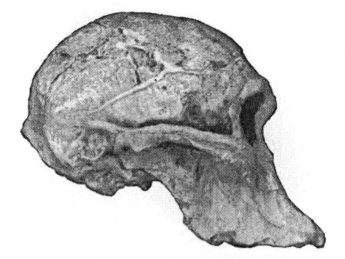

FIGURE 3.1 Skull of *Australopithecus africanus* from South Africa, dating to about 2.5 million years ago. These meek creatures set the stage for human evolution.

beyond the relative brain size of a modern chimpanzee. Whatever cunning they had, or whatever social organization they used, it was apparently sufficient to allow them to survive among the diverse crowd of animals that formerly inhabited Africa.

Nowhere is the rich biodiversity of these early times more evident than in the caves of the Makapansgat Valley, South Africa. The hills of the valley are spotted with caves, some of which are filling up today just like others that filled up long ago, only to be reopened by quarriers in search of limestone. One particularly large cave in-fill produced not only common cave formations such as stalactites, stalagmites, and flowstones for the quarriers but also preserved a rich cache of fossil bones. Interpretations of the fossils help paint a paleoenvironmental picture of a tropical forest some three million years ago, roughly contemporaneous with Lucy and her kin from some of the East African fossil sites. The mammal species represented in the Makapansgat cave fills were more numerous and diverse than those inhabiting the valley today.

Only a small percentage of the Makapansgat fossils are of *Australopithecus*. These were found in a conglomerate of broken bones that appear to have been accumulated by a prehistoric hyena. Alongside were numerous potential competitors of our ancestors, including at least six other primate species. Perhaps *Australopithecus* existed only in small numbers at the time; alternatively, these upright animals may have been savvy enough to avoid predation more often than their primate cousins. One thing is clear—*Australopithecus* was not a huge player in the ecological dynamics of the time. The tropical environment allowed the novel genus to slip into the diverse web of life with little impact. If *Australopithecus* was one of Darwin's "wedges," it was a small wedge hit lightly against the "yielding surface" of ancient Africa. Apparently *Australopithecus* served as an occasional meal more than as an ominous predator or competitor.

"Driven Inwards by Incessant Blows"

The successor to *Australopithecus* was a slightly brainier type of animal, usually referred to as the earliest member of our human genus, *Homo*. These creatures appeared on the African scene sometime around 2.5 million years ago. The earliest forms of *Homo* retained some of the features of their predecessors, including long, gangly arms that would have allowed them to climb trees when not walking on the ground. Nevertheless, they were clever enough to invent and use the first stone tool technology, simple flakes of sharp stone known as Oldowan tools (fig. 3.2). Despite the simplicity of the innovation, such tools were evidently useful enough to cut meat from scavenged carcasses. The diet of these early toolmakers certainly remained mostly vegetarian, but if they could quickly cut meat from some other animal's prey, they could get some easy protein to fuel their expanded brains. In ecological terms,

FIGURE 3.2 Oldowan tools were simple, unmodified flakes knocked off rocks. Despite their crudeness, the tools helped early *Homo* to "carve out" a new niche that included the scavenging of animal remains. (*Illustration by Michael Masters*)

we say that they entered a new "niche," or environmental role, when they began to supplement their diets with meat.

Don't think it was easy for early *Homo* to enter a scavenging niche, even with stone tools. Our ancestors had to contend with some fierce competitors. Even today there are many scavengers in Africa that give us cause for concern and caution in the wild. Jackals, hyenas, and vultures come immediately to mind, and back then each of these creatures had evolved pretty much into its modern form. But the ancient African scene included a greater variety of hyenas, including some rather large ones. When I hold their fossil teeth, I can almost feel the fright that these large hyenas must have instilled in their prey. The same is true for the plethora of predators who might have been more interested in making an easy meal of early *Homo* than in chasing them away and eating the carcass they were scavenging. Again, the modern predators — cheetah, leopard, lion — were there, but so were types of large saber-toothed cats. It wouldn't have been easy to run away on two feet.

How did early *Homo* survive the scavenging niche? We may never know for sure. The stone-tool cut marks on animal bones tell us what they were doing to get meat, but they tell us little about how our ancestors competed for a carcass yet managed to escape predation. Mind you, the *Homo* fossils with leopard tooth marks demonstrate to us that they did not always escape! I like to think of early *Homo* as using their upright stance in two ways for protection. One way, as noted early on by Charles Darwin, would have been to hurl stones at the oncoming menace. After all, their hands were free to do so. Another would have been to stand up and look as big as possible. Today one can frighten off certain predators by facing them, ripping one's shirt open, and holding the two sides out to look larger and more threatening than our scrawny bodies would normally suggest. Whereas early *Homo* presumably had no clothing, the same principle would apply.

But one small individual throwing stones or going through antics reminiscent of my boys' attempts to impersonate monsters would have had little effect on a pack of hyenas. Nor would two or three people, if one could call them that, present much of a threat. But imagine ten or twenty of these odd, upright creatures throwing stones and pretending to be more monstrous than they really were. There would have been safety in numbers, at least in this particular case. Many other primates, such as the baboons I'm so fond of watching in the wilds of Africa, use their troop structure for protection. The same would have been true for these incipient humans.

For over half a million years early *Homo* survived the African landscape, as did most of the creatures with which they competed for survival. Early *Homo* played the scene, but did not dominate it. Things changed dramatically around the beginning of the Pleistocene epoch, about 1.8 million years ago, both for our ancestors and for the animals around them. At that time, our ancestral lineage converged on a new form known as *Homo erectus* (sometimes called *Homo ergaster*). The best evidence for this evolutionary breakthrough comes from a fossil site called Nariokotome, west of Lake Turkana in Kenya. Alan Walker, now of Pennsylvania State University, described a magnificent find of a partial skeleton of a strapping young "man" (fig 3.3). The analysis shows that had this boy made it to manhood (again if one could call his kind man), he would have been about six feet tall.[3] Along with modern bodily proportions of this species came a further increase in brain size. Though the brain had not yet reached its modern capacity, our ancestors were poised to use their intellect for greater adaptability. The imposing figures of *Homo erectus* cast their shadows across Africa for more than a million years, expanding their niche and their geographic range.

Something interesting happened after the appearance of *Homo erectus*: other African mammal species started to go extinct at an unprecedented rate. Although it is difficult to tell from the fossil record exactly when a species went extinct, two separate analyses reveal a distinct trend toward a reduction in African species biodiversity once *Homo erectus* appeared on the scene. The human wedge was taking shape.

A team from the Smithsonian Institution led by Anna Kay Behrensmeyer found that between three and two million years ago, species biodiversity was on the rise in East Africa.[4] This may seem odd, since it was a time of global climate change toward cooler, drier environments—and species diversity seems to thrive in warmer, wetter parts of the world. But, it turns out, the increase in species biodiversity is an artifact, largely a product of the biases inherent in the fossil record. Basically, as we get closer to the present, we find that more and more fossils are well preserved; older fossils tend to disappear with time, and become less frequent in the fossil record (save a few extraordinary fossilization events). Despite that tendency to find more recent fossils than older ones (i.e., before they succumb to the ravages of time), the trend toward finding greater species diversity reverses itself sometime after two million years ago. Or so seems to have been the story in East Africa.

In order to confirm that the mammalian extinction rate increased around 1.8 million years ago, I took a somewhat different approach. Because all we have is a sampling of fossils at various time periods, I filled in the gaps using

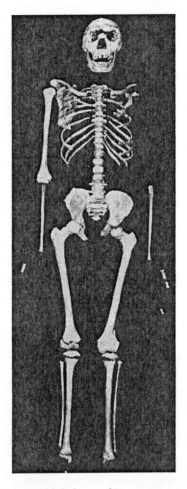

FIGURE 3.3 This 1.8-million-year-old skeleton of *Homo erectus* was discovered and exca-
vated at Nariokotome, Kenya. The appearance of this species accompanied the beginning
of a trend toward increased mammalian extinctions. *(Photo by Alan Walker, © National Muse-
ums of Kenya. Printed with permission.)*

computer simulations.[5] In short, I used my computer to create a hypothetical
community of mammals and, in the blink of an eye, have them evolve and go
extinct over millions of years. Once I had a simulated pattern of evolution,
the computer sampled the past much in the same way a paleontologist samples
the fossil record — it essentially dug up extinct species from the database. In
this way I could predict what we *should* find in the fossil record under various
scenarios.

What my computer and I found was that the past two million years of East

African prehistory involved a net loss of mammalian species biodiversity. There was no other way to account for the patterns in the fossil record. It was difficult to pin down the precise timing—after all, given the chance nature of fossilization, a species could have persisted long after one of its members committed its bones to the fossil record. But certainly we know that many of the species alive two million years ago are extinct today, and they appear to have started disappearing around the time *Homo erectus* was invested with the mantle of human evolution.

It is impossible to say what led to the increased rate of extinction. Usually environmental changes such as climate change precipitate extinction trends, but no dramatic shifts in climate are noted for that period. Indeed, the species that lived in Africa at the time had already survived considerable climatic upheaval.

There are other types of environmental changes that could account for the changes in the ancient African community of mammals. Sometimes the origin of a new competitor on the scene can affect the livelihood of another species. Given the complex web of life, and the multiple interactions of various species, even a modest new entry into the web can have a cascade effect and upset nature's supposed "balance." For example, today we have seen how the introduction of a single species can upset an entire ecosystem. I can witness that in my backyard, where European starlings are the most frequent visitors, outcompeting native birds for the food resources I provide. In 1890 a few European starlings were released in New York's Central Park, and now they are the most numerous bird species in the country.[6] Likewise, the crabgrass my neighbors and I struggle with was originally introduced to America as a grain. But I'm getting ahead of myself by looking at ecosystem shifts caused by human introductions. The point is that small entries can grow into big ones. So it likely was in the past, and over hundreds of thousands of years there should have been many dramatic shifts in the construction of life's web.

Could it have been *Homo erectus* who started the cascade of extinctions? It seems a remarkable coincidence that the declines in biodiversity correspond to the evolutionary progression of this species, though coincidences do happen: one could just as easily blame the downward trend of biodiversity on the evolution of a new species of warthog. Furthermore, the stone tool technology had not changed significantly. These ancestors of ours were still using simple Oldowan tools. Whereas we cannot blame the technology, we may be able to condemn the users of the technology.

One thing we can infer about *Homo erectus* is that their population size was growing. It is impossible to say how many of them there were, or how fast they were increasing their numbers. But the inference becomes obvious because of the geographic spread of their population. We find *Homo erectus* fossils across Africa. No longer do we have just scattered species of *Australopithecus*, but a widely spread species along the Rift Valley of East Africa, down to South Africa. As time went by in the Pleistocene, this species eventually spread into North Africa, and beyond. Indeed, *Homo erectus* was the first of our ancestors to move outside the confines of the African continent, possibly as early as 1.7 million years ago, as told by the earliest non-African *Homo* fossils — those of Dmanisi, Georgia.[7] Certainly by one million years ago, if not earlier, they had spread as far as East Asia.

The diffusion of *Homo erectus* into new territories could be explained without recourse to population growth. It might have happened simply by migrations of small groups. But as they had no destination in mind, and as Africa continued to serve as a bountiful habitat for them, it is more logical to propose that their populations were growing. It is worth conducting a quick mental exercise to see how this could happen. Let's start with an initial population of just twenty individuals, such as the clan or troop of the Nariokotome boy. At a population growth rate of just 0.01 percent (which is a small fraction of today's human growth rate of roughly 1.2 percent), there would be over 65 million *Homo erectus* individuals in just 150,000 years. (That is just a little more than a tenth of the current population of Africa.) If each person needed one square kilometer to find sufficient sustenance, then some of the growing population would have had to spread from Africa after 142,320 years. Keep in mind as well that not all of Africa would have been all that inviting: the Sahara desert was established by the Pleistocene epoch, and probably would have limited human expansion in that area. Moreover, we don't know how well our ancestors would have fared in other difficult environments such as mountains, which may also have impeded even the most adaptable mammal of the time. The tropical and subtropical regions of Asia would have looked more appealing to our ancestors, and thus we find the earliest evidence of movement beyond Africa at the West Asian Dmanisi fossil site of 1.7 million years ago — shortly after the appearance of *Homo erectus*.

It is evident that a very small growth rate can fill up a continent fairly quickly in evolutionary time. The growing prehuman population would have been competing with other mammals for many resources. It thus appears that the human wedge was driving itself into the African surface, and may have played a role in the increased rate of extinction among large mammals.

Granted, we do not know and cannot know the prehuman population growth rate from 1.8 million years ago. Nor do we have any proof that this allegedly growing population had anything to do with the mammalian extinctions. Neither assertion would hold up in a court of law, for the evidence is all circumstantial. Barring proof, which is hard to come by, can we establish guilt beyond reasonable doubt? To do so we would have to deduce a motive and look for evidence of a modus operandi—an "m o" that indicates a pattern of "criminal" offenses.

The Destructive Trail of Evidence

The question of motive is easily resolved—it was simply a matter of survival. These ancestors of ours had to eat the same fruits and berries as did the baboons and other monkeys. They were evidently eating the meat that other scavengers wanted, and it is not unrealistic to suggest that they may have consumed some meat that other predators needed. But *Homo erectus* need not have killed a single animal in order to lead to the extinction of others. They just had to get to the food resources first. If they were good enough at food procurement, the others would go hungry.

Evidence in support of the competition scenario comes from an irregularity not only in the rate of extinction but in the pattern of extinctions witnessed. It was not a random assortment of animals that went extinct, but those that somehow seem to have had particular relations with *Homo erectus*. The record of extinction is skewed toward reduced biodiversity in baboons, who were likely competitors for portions of their diet, and large carnivore species, who *may* have been competitors.

The extinction of large carnivores—extremely sizable hyenas and immense saber-toothed cats—is not so easy to pin on *Homo erectus*, at least in terms of competition for meat. Even though we know our ancestors were eating meat, it is not clear whether or not they actually did any hunting or if meat eating was a large part of their diet. But just as the skulls and teeth of these extinct beasts inspire awe today, they probably invoked a significant reaction in the past as well. The large carnivores may have been more of a threat than competition. So if the motive was survival, protection from predation would have been a significant concern for *Homo erectus*.

One may suspect that a bunch of prehumans tossing rocks at saber-toothed cats would not have put much of dent in the predator's population. But what if *Homo erectus* became a sly predator? What comes to mind is one of the most horrific and fascinating stories I've ever read about chimpanzee behavior. Chimps eat mostly vegetation, and occasionally are predators of monkeys and small game. But primatologists Richard and Jennifer Byrne witnessed a truly unique kind of kill conducted by chimpanzees of the Mahale National Park, near the eastern shores of Lake Tanganyika.[8] A group of chimps, exhibiting as much fear as bravado, were poised near the entrance of a small, narrow cave. From within that cave emanated the distinctive sounds of a leopard, giving particularly deep roars, audible to the anxious chimps — and the nearby Byrnes. The chimps would periodically enter, only to retreat with excited screams and barks.

Why would the chimps endanger themselves in a leopard's lair? After all, we know from the *Australopithecus* fossil sites that the chimps' cousins occasionally served as leopard prey (as do some modern humans and chimps). The answer came after about an hour of repeated charges into the cave, when an elder male chimp emerged from the cave toting a small leopard cub. The cub was still alive, but soon became battered and dying as the chimps repeatedly prodded it, lifted it, and dropped the cub on the ground. Rather than eat the cub, the chimps toyed with the dying body until they dragged the lifeless carcass away.

One can only speculate that behavior like this may have characterized our ancestors, but it serves as an example of the types of behavior that can accompany the motive of survival. The killing of a predator's young is an effective way to ensure relative safety in the future.

Survival is just one of the motives behind our case. We also have to find a motive for the alleged expansion of the population. There the motive is a bit easier to identify: sex. Like all mammals, we are motivated by our biology to perform acts that lead to reproduction. Reproductive success, accompanied by survival, simply leads to population growth. It need not be any more complicated than that.

Having established some circumstantial evidence and motives to build our case that *Homo erectus* population growth led to the demise of other species, we are left with the task of finding a pattern of behavior, or an "m o," that leads us to a verdict. And for that we have a world of evidence.

Pleistocene *Modus Operandi*

The fossil record can be analyzed on many different levels of biodiversity. So far we have been dealing with species biodiversity and its decline in Africa during the Pleistocene. Unfortunately we cannot look at past genetic biodiversity, at least not with our current analytical tools. But we can go up one level and look at the diversity of genera (the plural of genus). What is a genus? It is the first part of the species name, so *Homo* is our genus, and *Australopithecus* is the extinct genus of our early ancestors. The genus represents a higher level of categorization, encompassing a number of closely related species.

One might think that *Australopithecus* went extinct with the evolutionary origin of *Homo*, but only certain species of that genus went extinct (probably through a transitional extinction). Other species, the robust forms of *Australopithecus*, persisted for some time and lived side by side with our ancestors. The term "robust" refers to the sizable grinding teeth and thick-boned faces of these evolutionary cousins of ours. Their teeth and faces appear to have been well adapted to pulverizing hard vegetation — seeds and nuts and other prehistoric health foods. Thus they filled a separate niche from *Homo*, who was eating more soft fruits and meats. Nevertheless, the last of the *Australopithecus* fossils appear in the fossil record between 1 and 1.4 million years ago, after which the entire genus went extinct.

The point here is that the extinction of an entire genus is a big thing in Darwin's world. And genera can be counted like species, perhaps with greater reliability. Richard Klein of Stanford University did such a count of extinct genera in various time periods in various parts of the world.[9] In short, he found a telltale relationship between the number of extinct genera and the arrival of *Homo* on a continent. He found the modus operandi we needed to place blame.

Some of Klein's data come from South Africa, where he has worked for many years at a variety of fossil sites. Between 3.5 and 1.8 million years ago, during the reign of *Australopithecus* (of the nonrobust form) and early *Homo*, there was nothing peculiar about the extinction rate of genera. Indeed, the extinction rate may have been a bit low. Just as I found with my species data, our earliest hominid ancestors had not yet comprised a significant wedge on the African landscape. Between 1.8 and 1.0 million years ago, as the *Homo erectus* wedge apparently grew, the generic extinction rate went up dramatically (fig. 3.4). After accounting for the different lengths of time in the periods being compared, the proportion of extinct genera in the fossil record more than doubled (as compared with the number of all genera of the time period). Clearly something dramatic was happening in South Africa in the presence of *Homo erectus*.

Klein's data thus far have only confirmed the extinction trends we looked

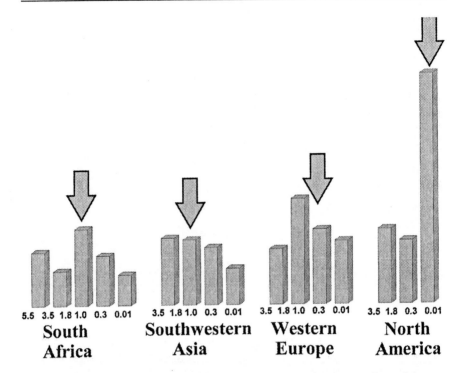

5.5 3.5 1.8 1.0 0.3 0.01	3.5 1.8 1.0 0.3 0.01	3.5 1.8 1.0 0.3 0.01	3.5 1.8 0.3 0.01
South Africa	**Southwestern Asia**	**Western Europe**	**North America**

FIGURE 3.4 Each stack represents the relative number of extinct genera for each region over intervals of time (in millions of years). The arrows show the times of entry of the genus *Homo*, and a pattern of increased extinctions that accompanied each grand entrance.

at earlier in this chapter. But as the investigation reached beyond Africa, the pattern became more clear. In southwestern Asia the data do not show a clear trend with the introduction of our ancestors, because of a smaller sample size, but the relatively high number of extinctions after 1.8 million years ago is consistent with the hypothesis that *Homo erectus* was having an impact. In western Europe the pattern is intriguing. We do not know exactly when *Homo* entered Europe, but certainly they were there by 700,000 to 800,000 years ago — and the high proportion of extinct genera around that time attests to this entry.[10]

But wait a minute. We're talking about the Pleistocene epoch, and wasn't that the time of the ice ages? Might considerable global cooling and the spread of ice sheets provide a better explanation for the extinctions than the impact of one species? Indeed, I've long argued that climatic change is more likely to lead to extinction than to evolution, and that would result in a decline in biodiversity.[11] But the pattern of climatic change started about

2.5 million years ago, and the most dramatic changes began only about 900,000 years ago. So the timing is off for a climatic explanation.

To test the idea that climate change, and not *Homo erectus*, determined the rate of generic extinctions, we need a continent that did not have the human or prehuman influence. North America provides just such a test. On that continent the extinction of genera did not show any rise during the ice ages (fig. 3.4). But after the end of the last glacial phase, when *Homo sapiens* entered the continent, the extinctions skyrocketed.

Is our case proved? Not by any means, but the evidence seems to show both motive and opportunity, as well as a pattern of behavior. And, once again, if we look at the pattern of extinction, we get some additional clues.

Summer in China and Springtime on the Riviera

A short drive from Beijing, China, one can visit "Dragon Bone Hill." This hill is riddled with cave sites and ancient cave in-fills. The fossils that were found there long ago were once thought to have belonged to dragons, and the local people ground these ancient teeth and bones into powders for medicinal purposes. In the early 1900s, a visiting German paleontologist found a tooth in a Beijing drugstore, and recognized it as that of a human rather than a dragon. This discovery was traced to its origin, and eventually led to some magnificent fossil discoveries at Dragon Bone Hill, a site now known as Zhoukoudian. Among the finds was a collection of skulls and bones of what became popularly known as "Peking man," the Asian form of *Homo erectus*.

Zhoukoudian is a remarkable site. In the summer, it has the same feel as the Makapansgat caves of South Africa; indeed these ancient caves formed in a similar kind of limestone. When I visited, the heat and humidity produced a smell that immediately took me back to my favorite haunts far away. But unlike the South African cave sites, in the winter the Chinese site may become blanketed with snow. And half a million years ago, when these caves started filling with the bones of our ancestors, the climatic regime would not have been much different. *Homo erectus* had extended the human habitat into the temperate zones and again expanded the human niche.

Along with the ancestral human bones at Zhoukoudian were the bones of many other animals who had been hunted by *Homo erectus*. Moreover, it is clear that fire was being used on a regular basis, a technological advance that clearly enhanced the hunters' occupation of this temperate region.[12]

Homo erectus evidently hunted a wide range of animals, including "big game." Not all of the bones in the cave deposits can be attributed to their hunting, for other carnivorous beasts also occupied the caves long ago, including a malevolent hyena that may be responsible for bringing in some of the human remains.[13] What is relevant to our discussion is the number of species found that seem out of place — species that are now extinct in China.

When one thinks of China today, one does not envision a mammalian community of bison, buffalo, woolly rhinoceroses, elephants, saber-toothed tigers, leopards, cheetahs, or striped hyenas. To be sure, many of the mammals found in the cave deposits have modern-day counterparts, but it is the pattern of extinction that leads one to suspect that our ancestors had a hand in the demise of both prey and competitors. One key reason for this suspicion is that once humans started spreading to Europe as well, many of the same kinds of species went extinct.

Along the French Riviera is an open-air site dating to about 300,000 years ago called Terra Amata. Whether it was occupied by *Homo erectus* or a subsequent species closer to us is unclear, as the only direct evidence we have of the humans who occupied the area is a footprint and some coprolites (fossilized feces). Nevertheless, clearly the occupants had considerable savvy, for evidence from postholes in the ground demonstrates that they constructed large shelters capable of housing up to twenty people. At the center of each of these oblong huts, which were probably covered with animal skins, was a fire hearth with human debris scattered around. And those coprolites, as unseemly as they may sound, reveal that the folk who lived at Terra Amata enjoyed the Riviera during the springtime (based on the springtime pollens that had passed through our ancestors' digestive systems). But again, it is the assortment of bones at the site that is of greatest relevance.

The bones at Terra Amata give further evidence of "big-game" hunting. The excavators found elephant, rhinoceros, and wild boar remains. The latter would not be astonishing to readers of the French cartoon series "Asterix and Obelix," two comic characters who have a fondness for feasts of wild boar. But elephants and rhinos do not fit our vision of modern-day France. They are now extinct there, and have been for some time. Other sites across Europe show that such animals were hunted and slaughtered in a most wasteful manner — with entire herds driven off cliffs or into swamps.

It seems curious to many that very large mammals such as elephants and rhinos survived on the African continent while being exterminated elsewhere. Indeed, Africa is often referred to as the "living Pleistocene" because such mammals persist there to today (although they are currently endangered in many countries). I hasten to note that Africa is hardly a "living Pleistocene" for the vast number of other mammals that went extinct, as we have seen. Even many large African mammals met their demise: the giant horse, the giant warthog, and the long-horned buffalo, to name a few.[14] But the survival of other large mammals in Africa needs some explanation.

One explanation of the persistence of African "megafauna," as such large beasts are known, is that humans originated in Africa and only gradually acquired tools of the hunting trade. Thus elephants, hippos, rhinos, and the lot had more time to adapt to the destructive human creatures. There may be an element of truth to this. As we'll see throughout this book, the quick entry of a particularly effective predator or competitor is often more disruptive and sometimes devastating to a community than the gradual spread of a species into an ecosystem.

One must also remember, however, that the later Pleistocene of Europe saw repeated glacial phases that no doubt placed hardships on a variety of species inhabiting the land. On the other hand, North America also had a variety of megafauna and other mammal species that survived each glacial episode. Their extinctions came only near the end of the last glacial phase, coincidentally when *Homo sapiens* spread across the continent. And therein lies another intriguing story to build our case against the human wedge, replete with the "smoking gun" evidence we require for a verdict.

A Mammoth Task

Outside of city zoos, North America does not have any extremely large mammals — no "megafauna." But it once did. Mammoths and mastodons roamed the continent during the Pleistocene. Other large mammals were plentiful, along with an assortment of large carnivores including the saber-toothed cats. When I sit at my river "office," I often try to envision the valley with beasts large enough to shake the ground or thundering herds of bison. All such creatures are gone from here now, and many of them went extinct across the continent in a geological flash of time, some eleven thousand years ago. Suspiciously, that is shortly after the first Native American people entered North America, armed with a hunting technology sufficient to fell just about any beast.

The mammalian extinctions at the end of the Pleistocene were so swift, selective, and profound that it is difficult *not* to point an accusing finger at the human predators. Nearly two dozen archaeological sites reveal a scatter of megafaunal bones in association with the sophisticated spear points that were the signature artifact of a culture known as Clovis (fig. 3.5). Yet reconstructing past events has never been easy. Controversy and intense academic debates surround our vision of the human entry into the New World and the subsequent mass extinction at the end of the Pleistocene, for indeed the extinction also coincided with the termination of the last glacial phase. Some scholars see rapid climatic warming as being more disruptive to the mammalian communities than a handful of Clovis hunters wielding spears. Others suggest that the ecological disruption was accentuated by the spread of disease among the animals. None of these perspectives on the past are nonsensical, and the extinctions may have been a product of many combined factors. For our purposes here, we must ask if the human factor was significant. Do we really have a "smoking gun" proof of the case, or, more appropriately, any "flying spear" evidence for the decimation of American biodiversity by the first people who entered the continent?

The controversy starts with a time period leading up to the extinctions. When and how did humans first enter the wilderness of the New World? The classic version is that wandering hunters from Siberia began entering North America around 15,500 years ago, following game across a land bridge called

FIGURE 3.5 Did these finger-length Clovis spear-points bring down the mighty mammoths and mastodons and precipitate a mass extinction in North America?

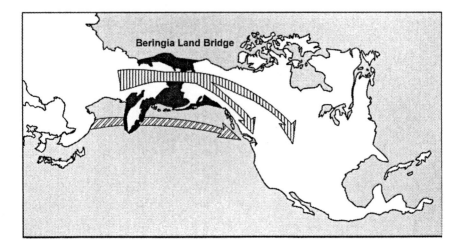

FIGURE 3.6 People probably first entered North America from Asia via the Beringia Land Bridge exposed during the last glacial phase, as they followed game into a new land. They may have traveled down the coast or through an inland corridor created as the massive glaciers melted. Alternatively, the first Americans may have arrived by watercraft following sea currents.

Beringia, in the vicinity of the Bering Strait (fig. 3.6). The sea level would have been low at the time, as much of earth's water was tied up in the ice of the massive glaciers that covered much (but not all) of the northern half of the continent. The lower water level exposed the land bridge, and humans could have sauntered across and followed routes along the coast or through an ice-free corridor that appears to have bisected the glaciers.

Given the opportunity of entry 15,500 years ago (or earlier), it is not surprising that one of the earliest human occupation sites found by archaeologists in the Americas dates to 14,700 years ago. What *is* unexpected is that this creekside campsite, Monte Verde, is in southern Chile—twelve thousand miles from the alleged point of entry. This requires some explanation.

It seems unlikely that wandering peoples would traverse such a great distance from Beringia, over a wide variety of terrains and through diverse ecosystems, and end up at Monte Verde in under one thousand years. This leaves at least two possible alternatives: they got there by boat, or people entered the New World by foot considerably earlier than we thought. The boat hypothesis is not as ludicrous as it sounds—after all, the earliest inhabitants of Australia got to that continent over the water about 56,000 years ago (and possibly earlier). On the other hand, the Pacific is a vast and difficult ocean to cross, unless the voyage happened to cross it much farther north than Chile at an earlier time and/or proceeded down along the coast by boat. Thus there is growing yet cautious acceptance of the notion that the first Americans may have entered long before the era of the Clovis hunters. When and how, we are not sure.

Of this we can be certain: humans were in many parts of the Americas by the end of the Pleistocene, and encountered a wide variety of creatures that had no specific adaptations to human predators. This is quite different from the African situation, where hunting humans and modern megafauna evolved together. But how destructive could the Clovis people really have been? Yes, we do have archaeological evidence that they hunted mammoths and mastodons, but the small number of sites found so far does not suggest that humans were overwhelming the continent. On the other hand, it is worth once again imagining some population scenarios.

If the first human entrants to North America arrived 15,500 years ago, it is true that they would have encountered a hunter's paradise of "naive" fauna — those with no experience or adaptations to fend off human hunters. With no shortage of protein for their diets, and sufficient plant foods as well, there is no doubt that their populations grew. Early Clovis sites are found from Alaska down to the southwestern and midwestern reaches of North America, and one can be certain that they were not inhabited by just one small group moving from place to place. Just for the sake of the argument, let's start with an initial population of 100 people trickling across the land bridge, and allow their numbers to grow at a modest rate of 0.5 percent per year. (Once again, this is just a fraction of today's human population growth rate.) After the first 3,000 years, leading up to the time of the mass extinction, there would have been nearly 315 million people. This seems unlikely, as that number is greater than the current population of North America (most of whom were products of population growth of the past hundred years). It simply illustrates the point that even with a slow growth rate, 3000 years can see a significant rise in population numbers.

To get a more realistic figure, we can try to factor in some other variables. As before, we can give each individual one square kilometer, and see what kind of growth rate would have led to saturation of the North American landmass in just 3,000 years. It works out to 0.42 percent per year. Growth rates less than that show considerably less impact; such is the nature of population mathematics. So, for example, at a slightly slower annual growth rate of just 0.4 percent, the Clovis people would have covered roughly 65 percent of the North American land mass. Keep in mind that the glaciers had not fully receded, which would have pushed the mobile populations southward. Thus the continent easily would have had much more than a handful of hunters.

An alternative scenario is tempting, especially given the early date of Monte Verde way down in South America. If there were pre-Clovis peoples, the population growth could have started much earlier. We could go back to a time when the Beringia Land Bridge opened for a lengthy period starting around 35,000 years ago, but there is no good evidence for any human migrations into the region at that point—the earliest archaeological sites in Siberia date to sometime after 20,000 years ago.[15] If we take a more conservative guess of 20,000 years ago, with some fishing groups boating across from northern Asia and settling in North America, our scenario gains some interesting dimensions. This early entry is not unreasonable; remember the earlier Australian entry over water, and it is clear that the technology would have been available for water travel. Moreover, the northern reaches of Asia and North America would have been enjoying milder winters than those characterizing the height of the glacial phases.

Starting with the same simple assumption of 100 initial people, we can work out what would happen over time under various growth rates. But now we have 9,000 years to deal with before the major extinction event. Saturation of the continent (still assuming one square kilometer each) would occur with an annual population growth rate of just 0.14 percent. That number may seem fairly arbitrary and obscure, so it is worth fleshing out some of the details such a growth rate entails. It means that a population of 100 people would take about eight years to become 101. Some people might die in those eight years, some might be born. The excess after eight years would be just one person. That is not a lot of growth, even in a foraging society. But with time it can fill a continent with people, and put their prey and competitors at risk.

Playing with the numbers does not give us a final resolution to our problem of relating human population growth to mammalian extinction. It only shows what *could* have happened, and gives support to the notion that there may have been sufficient hunters to devastate populations of their prey. We must acknowledge other possibilities, and take a look at the possible effects of climate change. Could the human factor have been insignificant when compared with a rapid retreat of the glaciers some eleven thousand years ago?

Global Warming and Extinction

It was not just huge prey animals that went extinct eleven thousand years ago, for some small animals were lost as well. However, most of the extinctions were of animals that weighed more than one hundred pounds. As many as thirty-four genera of such mammals went terminally extinct in North America, along with three genera of small mammals.[16] In South America fifty-one genera went extinct as well, only eleven of which were shared with North America. Included among the victims were mammoths, mastodons, saber-toothed cats, horses (of the native American type), large rodents, and a bewildering array of odd-looking creatures.

This variety of mammalian extinctions has suggested to some scholars that climate change may have been the cause. It does not seem likely that humans were hunting all of these species down to the last individual. Moreover, the warming trend was quite rapid. Keep in mind my mantra that climate change, and particularly rapid climate change, does not give species time to adapt, and that extinction is a more likely outcome than evolution of forms sufficiently modified to survive the new environmental regime. In the biological game of adapt or die, the latter option seems to have been more common over the past few million years.

So why did so many mammals go extinct? Most of them had survived a number of glacial and warming cycles. Perhaps the speed of the change made the difference this time, and not the entry and expansion of the human populations. Many of the mammals would not have been as resistant to climatic change as they were resilient. What's the difference between resistance and resilience? Resistant animals are the adaptable ones who can maintain large populations throughout a variety of climatic regimes. Resilience works another way: population levels for resilient species decline during unfavorable climates as their preferred habitats shrink, and then bounce back with rapid population growth once the climatic cycles allow the reestablishment of the species' preferred habitats.

Large animals tend to have slow reproductive rates due to long periods of gestation and individual development to adulthood. They also have smaller litters. Their populations do not bounce back as easily as those of an adaptable animal such as humans; witness the population growth rates discussed earlier. Thus the devastation that comes with swift climate change can lead to the decline of a species that previously showed resilience when the change occurred more gradually.

A further argument against human impact is that some large mammals *did* survive, despite being human prey. And, as the human population continued to grow for the next few thousand years, the high extinction rate did not continue. The extinctions are mostly focused around eleven thousand years ago. When the rate of climate changed slowed, and things warmed up toward today's temperature levels, fewer species went extinct.

What is naggingly curious, however, is the pattern of extinction. Whereas many creatures went extinct, again the majority were large land mammals. Large marine mammals did not go extinct. Nor did most small vertebrate animals, insects, or plants. In the face of rapid climatic change, some of these species *should* have gone extinct. So we are back to square one, and the conundrum explains the vehemence of the academic debate.

This much we *do* know: the mass extinction of large land mammals coincided with climatic change and the expansion of human hunters across the Americas. We also know that ecological systems are complex, and that large-scale events may have multiple causes. Could it just be that the devastation was a product of ecological changes wrought by *both* climate and humans?

Let's look at a few scenarios. If mammoths and mastodons alike, along with many other creatures, were more resilient than resistant throughout previous climatic events, then we can expect that their numbers may have been low. Humans, of course, have proved to be largely resistant to climatic and environmental change—we adapt with tools, clothing, shelter, fire, and social structures to move freely throughout almost any environment. Human numbers would not have stayed low, and we do know that they were hunting. Humans could have been the proverbial straw that broke the camel's back, or, more appropriately, the environmental change that sealed the fate of many large mammals.

Hunting of large mammals is also, by its very nature, presumably very wasteful. So, you kill a mammoth with a series of well-placed spear throws, and down it falls. The next trick is to get the skin and meat off it for clothes and food. This takes time, but soon your hunting party has a good deal to carry back to the camp. Imagine the feast! After all appetites are satisfied, one can assume that everybody has a good snooze—a lot of mammoth meat could produce quite a postprandial slump. But energized the next day, a team goes back to the kill site to gather more food and skins. Meanwhile, other creatures have taken interest in the carcass, ranging from vultures and wolves to bacteria and molds, with insects finding a nice warm place to lay their eggs. Within a very short period, the resultant stench would be enough to turn away most humans. It would be time to pursue the next animal for fresh meat.

Having nipped the resilience of the mammoth in this scenario, thus leading to the extinction of a mammoth species, what would have been the consequences? Would people merely have shifted to another naive and vulnerable species for their meat and skins? Possibly, but humans were not the only significant part of the mammoth habitat. Large mammals, it turns out, are very important parts of the natural ecosystems where they live. Their demise alone would have had a cascade of consequences.

A Mammoth Domino Effect

Probably the best person to ask about the consequences of large mammal extinctions is an ecologist named Norman Owen-Smith. During my years at the University of the Witwatersrand I had the pleasure of calling him colleague, and learned of his interest in the ecology of megaherbivores. Take, for example, the African white rhino and hippopotamus, two very large, grazing herbivores. They are basically lawn mowers when it comes to feeding: they graze the grass down to short "lawn" grasses, as opposed to the tall grasses one typically envisions on the African savanna. But it takes fire to maintain a tall-grass savanna; indeed, the tall-grass savannas that tourists enjoy are largely artificial products of human-induced fires, which promote grass growth and burn off tree saplings. The short grasses do not burn so easily, and thus fast-growing woody plants are able to invade and quickly transform the land into a thicket of dense woody vegetation.[17]

The ecological web associated with megaherbivores gets more involved. The disturbance of the land by the grazing of white rhinos and hippos as well as by elephants, who knock down larger trees, promotes the growth of woody plants that support other herbivores. The fast-growing woody plants have fewer defenses, are more productive, and have greater nutritional value for browsing antelope such as bushbuck or kudu. Thus the megaherbivores are "keystone" species in promoting and maintaining biodiversity of both species and ecosystems. Like the loss of a keystone at the crown of an arch, which would make the structure crumble, the demise of a keystone species can have a falling domino effect.

What happens if you take away the megaherbivores? You get either tall-grass savannas or forests, and lose much of the diverse vegetation preferred by many animals. The loss of one keystone species has a cascade of effects on others. Certainly a similar set of ecological interactions could have been relevant to the North American extinctions at the end of the Pleistocene. The grazing of mammoths or the browsing of mastodons helped maintain diverse ecosystems. With their demise the environments could only have altered dramatically, reducing the diversity of species as well as ecosystems.

Norman Owen-Smith makes yet another point that is relevant to our discussion. He notes that among white rhino populations of South Africa, areas with fewer rhinos have lower population growth rates.[18] He attributes this to one or both of two reasons: it is more difficult to find mates in smaller populations, and there are fewer rhino to "disturb" the vegetation in a way that promotes the growth of short grasses. This accentuates the point I made earlier, that many of the mammals put at risk by either human hunting or climatic changes would not have been resilient. Thus even a small human impact could have led to the demise of a large species.

The Smoking Gun?

Ecosystems are complicated entities — that much should be clear. Accounting for all the features in an ecosystem is thus a daunting task, especially if one is trying to pin a mass extinction to a single cause such as humans. We've seen that one way to deal with ecological complexity, at least in part, is with computer simulations. John Alroy, a paleontologist at the University of California, Santa Barbara, has created just that — a simulation of mammals of the late Pleistocene in North America including the possible impact of humans. Guess what? He thinks he found the smoking gun.

Alroy took many factors into consideration while creating this computer model, and varied the parameters for a series of simulations.[19] Such models can never take the entire complexity of life into account, but they allow those of us who use them to look at the effects of each parameter. Two of the important factors that Alroy programmed in were human hunting efficiency and the human population growth rate. Interestingly, these were tied together such that if hunting ability was exceptionally poor, then it was the human populations rather than those of the megaherbivores that crashed! But even with modest skill, the humans in his model survived — as we know they did from the archaeological record.

Here is what is important for our discussion: as long as the human populations grew, ecological devastation was inevitable.[20] The dates of extinction churned out by the model were governed largely by varying the assumptions about growth rates and population density. Moreover, the ecological changes suggested by Norman Owen-Smith's insights became evident. Wrote Alroy: "If accurate, these figures imply a major disruption of ecosystem function at the continental scale, with potentially severe consequences for vegetational structure, the size of vegetational carbon sinks, watershed dynamics, insect and small vertebrate population dynamics, and so on."[21] All this would happen with or without climate change.

We might not have proof, but the American scenario certainly fits the human modus operandi. A bit more evidence from another part of the world may convince the jury.

Species Going Under in the Land Down Under

Imagine the continent of Australia some 56,000 years ago, when the first humans got there. Some of the unique creatures we think of, such as the kangaroos or koala bears, had evolved their unique features on this isolated land and successfully made Australia their home. But there were other species that only the earliest human migrants may have seen. There was a much larger, claw-footed kangaroo to greet the new arrivals. The heaviest bird ever known was walking, but not flying, across the continent. It was gone by sometime around 46,000 years ago. Along with it went twenty-three of the twenty-four genera of large Australian land mammals.[22]

Another human entry to a new land, another mass extinction of large animals. Was it yet another mere coincidence? Can the extinction be attributed to other causes such as climate change? Not this time. It is true that Australia increased in aridity during the last glacial maximum, and such climatic change would have made it difficult for many large species to survive. But most were *already* extinct, and had been for as long as 20,000 years, predating the mass extinction in North America. The only logical cause of extinction was human intervention on the landscape.

Apparently hunting was not the sole cause of the rash of extinctions. The early Australian aborigines appear to have burned off much of the vegetation, which would have helped in hunting and traveling, but would also have wreaked ecological devastation for the large animals that consumed the vegetation.[23] Some of them may never have been viewed by human eyes before succumbing to the human impact.

It took some time after the initial human colonization before the extinctions took place. Was it because the new inhabitants were cautious? Probably not. More likely, it was because it took time for their populations to grow. The initial human migrant population must have been relatively small, having had to traverse a good bit of the sea to get to Australia. It would have taken considerable time before their numbers were substantial. But the extinctions took place after sufficient time, and were simultaneous in both eastern and western Australia.[24] If humans were the cause, then they must have spread out in great numbers.

Island Paradise?

On continents, it takes time for human populations to grow to the point where their impact can be felt on a broad ecological scale. Islands are much smaller, so the human-caused extinctions are more evident and much quicker. It is indisputable that humans were the direct cause of numerous extinctions on islands as widely dispersed as New Zealand, Madagascar, and the Galápagos. Hunting, habitat destruction, and human-introduced animals are the combined forces that led to the demise of many an island species.

New Zealand provides a particularly useful case study, for we can contrast it to neighboring Australia. The Polynesian settlement of New Zealand began just before A.D. 1300, and the immigrants soon started hunting large flightless birds known as moas. Does this scenario sound familiar? What is unfamiliar is the speed of their demise: by the best estimates, eleven species of moas were extinct within 160 years or less.

The estimated time of moa extinctions comes from a combination of archaeological evidence and a computer model devised by two New Zealand researchers, R. N. Holdaway and C. Jacomb.[25] As in so many other models we've explored in this chapter, they started the scenario with a colonization by 100 people, and let the population grow at a modest rate. Depending on whether or not habitat destruction was considered, and on the rate of population growth, the extinctions appeared to be inevitable — in less than one hundred years by some scenarios. Again, climate change was not a factor, for the moas had survived considerable climatic fluctuations before the arrival of humans.

Islands are small, and even slow rates of human population growth lead quickly to high human densities. These people have to eat and utilize resources. Their wedge, "driven inwards," pushes others off the brink. In the case of the moas, their inability to rebound from the onslaught was exacerbated by their slow rate of reproduction — they had no resilience.[26] The consequences could not be avoided, and I think it is fair to surmise that the hunters were oblivious to the extinctions they were causing, until it was too late.

An important common thread running through both continental and island models of extinction is human population growth. Even for those who hold to climatic explanations of the North American extinctions have to admit that human-induced extinctions become plausible, if not likely, once the population of people reaches a certain point of land saturation. It just happens more quickly on small islands, and is easier to see in retrospect. Thus our case against humans as the most significant wedge of extinction can be closed only if we know that their populations grew. And by all the archaeological evidence, certainly they did.

The only remaining puzzle is why extinction rates slow down while humans continue to increase their numbers. Here Norman Owen-Smith adds another insight: "Notably, humans are unique in being the only omnivore that is also an effective predator, enabling their populations to be maintained even when prey become scarce."[27] There are other things for humans to eat, and some of the prey animals reproduce quickly enough to survive the onslaught. Moreover, people are clever at garnering resources. We can maintain large and growing populations by growing our own food, domesticating plants and animals for our own purposes. It seems like a nice compromise, so that we don't have to exploit the wild, natural plant and animal resources with such reckless abandon.

But if people were oblivious to the extinctions caused by their hunting prowess, the extinctions caused by new techniques of subsistence would have been even harder to see. The reality is that agriculture is a more insidious form of environmental destruction. When we started to control the land, it was not the point at which extinctions ended, but the point where the biodiversity crisis really began.

4

C H A P T E R

Genesis of a Crisis

The introduction of agriculture and the ensuing rise of early civilizations are among humankind's greatest achievements. They have provided us with wondrous things and aided us in reaching our potential in many domains. Without food production, most of what we know as everyday life would not exist, and for that matter, most of us would not exist. Yet, it is also true that the evolution of a food-producing economy has taken a serious toll on humans and on our environment. By examining the development of agriculture, we may learn lessons for our own future.

CHARLES L. REDMAN, 1999

FORAGING was the human occupation for most of the time our species has spent here on earth. Growing our own food is a relatively recent innovation. Before that, generation after generation of our ancestors gathered fruits, nuts, and grains for the bulk of their sustenance, and wood for their fires. Animals were hunted for nutrition from the meat and marrow, as well as for skins to make clothing or shelter; horns and antlers were fashioned into tools. The lives of these foragers involved a lot of travel — every day they walked from their camps to find food and, where necessary, collect some water. As their vital resources dwindled, or as the seasons changed, they could pick up their few possessions and wander along to more hospitable lands.

Many of these foraging groups, perhaps clans or congregations of families, dotted the landscape. There was plenty for everybody, and as they moved from camp to camp, catchment to catchment, the places they had lived would replenish themselves. In the new season, the people would return to find buds on the trees, tassels on the grasses, and growing fawns, chicks, and calves. The plants and animals would have transformed the energy from the sun and minerals from the earth into calories and nutrients available for human consumption. Upon the peoples' return, more often than not they too would have added to their numbers with babies and young children. Their success at utilizing the land meant that their own populations were growing.

At different times in widely dispersed regions of the world, the human foraging strategy became restricted and eventually cramped.[1] As people moved to their favorite camps, they found other people had gotten there first, or that their favorite prey had been exhausted by others. Their competition for food and other resources came not so much from other species as from their own. One can only imagine what the elders of a family may have felt when they found other people on the land they thought was theirs. Thoughts must have ranged from the joy of finding new friends to anger at the intruders. Sometimes they cooperated, and shared ideas — and cultures grew. Other times, the worse aspects of our culture manifested themselves with greed and violence. Yet the human population continued to grow.

About ten thousand years ago, in the Middle East, the human way of life was about to change forever. Nowhere is this transition more evident than at the ancient "tell" of Jericho, in the Jordan Valley, just north of the modern city of Jericho. A tell is basically just a large mound of human debris that grew over many generations as newer structures were built upon older habitation sites that had crumbled to the ground. The sequences revealed in a tell are an archaeologist's dream, for each successive layer holds clues to the changes people saw in the past. Tell es Sultan, as the ancient site is known today, shows the changes wrought by the origins of agriculture.

The bottom layer of the Tell es Sultan contains remains of a settlement inhabited by foragers, before there was a city of Jericho. It was a small collection of homesteads, with a few circular house floors. Already, people had begun to congregate and establish long-lasting structures, even though they were used for only part of the year. Scattered about the ground were hunting tools and the remains of wild plants and animals — the last local evidence of a foraging lifeway that had served humanity for most of its existence.

Upon the foraging remains of Tell es Sultan was a considerably larger settlement, built about ten thousand years ago. Its remains represent what must have been one the world's earliest farming villages. The mud-brick houses were larger and more numerous, sufficient for about three hundred people. Ditches attest to the construction of waterways that diverted floodwaters from the people's homes. Walls and circular towers testify to the endeavors and cultural innovations of this settled population. And among the debris of the Jericho remains were preserved grains of domesticated barley and emmer wheat. People had taken nature into their own hands, and made it work for them.

These people still hunted gazelle, wild boar, cattle, and goats. We do not know if they had started herding any of these wild animals, as animal domestication was yet to come. Soon it did.

Across an area known as the Fertile Crescent, stretching from the Mediterranean to the Persian Gulf, many villages followed suit in the thousand years after the people of Jericho created a new way of life. Unlike the arid lands of today, after the end of the last ice age the Fertile Crescent was rich and ready for the seeds of the agricultural revolution to be sown. The plant and animal remains show telltale signs of a new force of evolution controlled, at least in part, by human initiative. And the human populations grew upon the bounty ensured by the work of their own hands.

The same process began later, probably independently, along the banks of the Yellow and Yangtze Rivers in what is now China (fig. 4.1). Later still, in South America along the Pacific slope of the Andes, in central Mexico, and midcontinental and southeastern North America, different domesticated crops and animals gradually appeared over the course of a few thousand years. In terms of geological time, it all happened in a blink of an eye. What was it that led to these innovations? How did it happen that widely dispersed peoples would suddenly find similar means of marshaling the forces of nature? Therein lies yet another topic of academic speculation and debate.

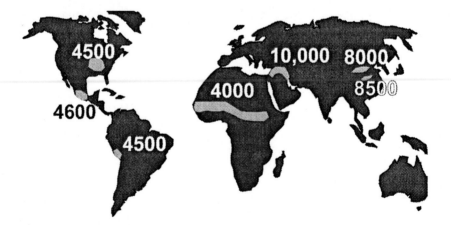

FIGURE 4.1 Agriculture arose independently in a number of places around the world between 10,000 and 4,000 years ago, starting in the Fertile Crescent of the Mideast.

Coevolution and Musical Chairs

With agriculture appearing independently across the world in a relatively short period of time, it would appear that there must have been a common cause. But seldom do complex phenomena have simple explanations, so identifying how and why agriculture arose becomes a difficult problem.

Environmental causation, as in so many arenas of assessing changes of the past, had its turn as a popular view for the origins of agriculture. Famed archaeologist V. Gordon Childe proposed that humans were pressed into a mutually dependent relationship with plants and animals when drought struck the Mideast.[2] The drought was seen to force humans and their potential domesticates into tighter quarters near water sources, and somehow agriculture developed out of this close association. Indeed it *is* true that the world had just emerged from the last ice age, and that the climate would have had an effect. But it was opposite what Childe proposed, for certain areas were environmentally primed for the initial stages of agriculture — hence the term Fertile Crescent.

Could it be that a more favorable climate spurred the origin of agriculture? There is no doubt that it was a necessary prerequisite, for climatic conditions during the recent glacial period were highly unstable and not conducive to long-term production.[3] But one is left to wonder why these events did not take place in a previous interglacial phase when equally clever people lived on comparably suitable lands. The archaeological record shows no evidence of agricultural tools or domesticates before the most recent glacial phase began. Thus climatic explanations, viewing humans as either forced by bad climates or inspired by good ones, suffer from the same fate as explanations of the extinctions at the end of the last glaciation. Something is missing from the explanation.

Another of the early ideas for the origins of agriculture came from Robert Braidwood, a pioneering archaeologist at the University of Chicago.[4] He envisioned a sort of cultural evolution, in which agriculture became possible for people who had reached a particular stage in their social development. If they lived in a suitable environment, then agriculture would have been inevitable; in earlier interglacial phases, culture was not yet ready for such an advance. This concept of a straight-line development through successive stages of cultural (or biological) evolution has long since been abandoned by most scholars.[5] Whereas cultures change and "evolve," they may do so in a myriad of ways. Moreover, the cultural evolution theory provides no explanation as to why agriculture happened so quickly or exactly how it happened.

The origins of agriculture involved not only human cultural evolution but also the evolution of plants and animals. The newly evolved life-forms, with humans contributing as agents of natural selection, are what we recognize in the archaeological record as the markers of domestication. So perhaps what we need is a theory of "coevolution." David Rindos of Cornell University provided the most comprehensive treatment of just such a theory.[6] Whereas not all scholars concur with Rindos, there is strong appeal in a co-evolutionary approach.

In essence, domesticated plants and animals initially evolved by utilizing humans as an important part of their environment. This is not unusual in nature — trees evolved fruits that attracted birds or monkeys, who would then happen to propagate the tree by eating the fruit and spreading its seeds. Likewise, grasses with tightly packed seeds in their tassels would attract people much more readily than those with widely distributed seeds that required more labor for harvest. Humans would unconsciously be the agents of natural selection by harvesting the most "fit" grains. This is what Rindos terms "incidental domestication." In effect, the evolving plants were the ultimate cause of agriculture.

As people settle into more permanent communities, they tend to disturb the land. Trees and plants are cleared to make room for homes. Paths are worn around homes and out to foraging areas by daily activity. Thus it was no accident that many of the initial domesticated crops were seed plants that readily took root in disturbed ground — essentially human activity had tilled the land and provided an inviting garden bed for barley, wheat, and lentils. Moreover, people preferred grains with thinner seed coats, as they were easier to process; they also happened to germinate more quickly.[7] Harvested seeds that would drop into the disturbed soil beds, or into the local garbage heap, would sprout before long as a conspicuous "weed." Eventually somebody had to notice the pattern, and take a cue from nature that there may be a better way to ensure an annual harvest. Meanwhile, the plants selected lost their ability to propagate naturally, and became dependent on the incipient farmers for dispersal. "Specialized domestication" (as contrasted with incidental) was born, and so was agriculture.

Domesticated animals probably evolved in ways comparable to that of plants. Dogs were the earliest domesticate, having evolved from wolves who found human hunters to be an ally. Later domesticates, such as goats or sheep, unwittingly "used" humans for the protection necessary to ensure their own breeding. Hence the first animals selected were social animals that could satisfactorily live together in confined places, and they quickly evolved to become more suitable to humans.

A coevolutionary model explains the "how" but not the "why" or the "when." Thus some elements of the explanation need filling in. For example, why did people settle in the first place? After all, the best way to ensure sufficient food in a foraging way of life is to keep on moving to the best areas. If you stay in one place, you are at risk when there is a bad year brought by drought or overexploitation of the resources. So we also need an explanation of sedentism. I hasten to add that not all agriculturists were sedentary, and not all settled people relied on agriculture, but the generalized nature of what I present here is meant to lead us back to the topics of this book: population and biodiversity.

According to archaeologist Michael Rosenberg of the University of Delaware, "both sedentism and agriculture are the products of population pressure."[8] Rosenberg cleverly compares a foraging way of life to a game of musical chairs. When the music plays, one gets up and moves to a new chair, just as seasonal change requires people to migrate to new environments. In the game, one chair is removed and somebody does not get a seat when the music stops; in real life, there are fewer and fewer nonoccupied places to go because of population pressure. Rosenberg clinches the analogy: "In a sense, sedentism can be described as cheating at musical chairs — refusing to get up when the (seasonal) music starts."[9] In real life the "chairs" are not equal — the territories vary in their productivity, so sedentism is a way to ensure that your family holds on to the best possible resources: you sit on the comfiest of chairs and don't budge.

Once a group is settled, its subsistence base must come from a smaller area. It is too difficult to "sit" on a wide expanse of land with all the resources a community needs and still manage to defend such a large territory from migrating foragers or other would-be settlers. Thus a group must either exploit resources more intensively — which could lead to overexploitation — or become more innovative in exploitation. Those who encountered the coevolutionary system envisioned by Rindos would have had an advantage, and would have been selected for success and community longevity. Others would have dwindled in number due to the lack of resources, or would have been pushed to marginal environments. Thus the only foragers left today, such as the Khoisan people of the Kalahari Desert in southern Africa, or the aboriginal peoples of Australia, live in places where agriculture has not been possible.

Do we know that populations were growing? Sometimes it is difficult to tell from the archaeological record. At sites such as ancient Jericho, the expansion of the community in terms of larger and more numerous mud-brick houses is evident. In other places it is not so clear. Rosenberg cautions that population pressure need not be equated with population growth; dwindling resources due to environmental change or overexploitation (or extinction of prey!) can also lead to population pressure. But in general, human populations tend to grow, and eventually population pressure becomes inevitable.

Fred and Wilma

Certainly there does appear to be a general correlation between population growth and the origins of agriculture. But correlation does not necessarily imply causation. Could it be that the origins of agriculture caused the populations to grow, and not vice versa? It is the classic chicken and the egg conundrum: which came first?[10] The notion that agriculture spurred population growth is common in the academic literature. But anthropologists who have studied the bones of early agricultural communities note something startling: the health and longevity of individuals declined with this new way of life. This was due to a number of factors, including greater focus on fewer and less nutritious food types as well as the increase in diseases that comes with tightly congregated peoples. This is hardly the stuff of a population boom. Yet the inexorable growth of the human population continued. It was part of a long-term process that gets at the heart of the timing of agricultural origins across the world.

In order to assess the relationship of population growth to the origins of agriculture, we need to conduct another thought experiment. It begins 100,000 years ago, when fully modern members of our species first appear in the fossil record. This time we'll start with just two people, the founding members of modern *Homo sapiens*. You may wish to think of them as Adam and Eve from the biblical book of Genesis, but I prefer to refer to them as the cartoon characters Fred and Wilma Flintstone — two distinctive members of a population. (Never mind that the dinosaurs in the Flintstones cartoon series were at least 65 million years out of place.) To be sure, their children would have bred with other individuals who were not quite yet of our current human form. But ultimately it was Fred's and Wilma's genes that initiated our species before we diverged into our modern variants. So for argument's sake, we can start with just two people.

Fossils do not reveal much about how quickly human populations grew. And certainly the rate of growth would have varied dramatically through time. For example, we know that past growth rates were not as great as today; had they been so, then it would have taken only 1,757 years to get to the size of today's population. But if one can imagine a steady climb from Fred and Wilma to the one billion people earth had before the industrial revolution, we can calculate the rate of growth. According to my computer, the population would go forth and multiply at a rate of 0.02033 percent for each of the 100,000 years.

Why would the rate be so low, when we know that human populations can and do grow at much higher rates? Well, it is a product of averaging out the experiences of many human populations over many years and across numerous geographic areas. No doubt during glacial times there would have been some populations that suffered losses or whose growth stagnated. Even in interglacial times, populations in more inviting environments may have grown at accelerated rates compared with those in the more northerly reaches of Europe and Asia. Growth rates fluctuate over time and space — in the next chapter we'll see that they still do — but the overall effect seen over the broad spectrum of time is one of fairly steady, relentless growth of the human population as a whole.

I've made a graph of this hypothetical steady population growth for figure 4.2. What I want you to notice is what happens around ten thousand years ago. Even at that slow rate of growth, the population numbers start to take off dramatically. I've seen similar graphs from archaeologists who use what little evidence we have to show that the human population grew after the origins of agriculture. Indeed it did! But the point is this: it did not grow because of agriculture but grew anyway, in some respects despite agriculture. There were more humans because there were more humans making more humans — that is simply the nature of population growth, even at the slow rate of 0.02033 percent per year.[11] It is simply the product of more and more people having children, and reaching a critical mass that had not been reached in earlier interglacial stages.

It is true that without agriculture the population may have stopped growing. There were simply not enough resources on earth to allow foraging populations to continue their way of life and increase their numbers. This is what happens to most species when they hit the carrying capacity — the maximum sustainable population — of their habitats. Either the population pressure has to be alleviated by a new mode of subsistence, or the growth has to

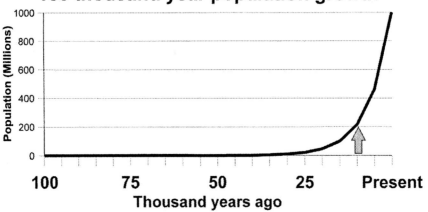

FIGURE 4.2 Even at a slow and steady rate of population growth, the human population would have taken off dramatically around the time of the origin of agriculture (at the arrow). Such is the nature of exponential growth.

stop. Being that the early agriculturists were probably unaware of the growth problem, there was no incentive to stop having babies. Indeed, children could be used to tend the fields and the herds of domesticates. What ensued was a positive feedback loop, or autocatalytic process: population growth spurred the origins of agriculture; agriculture fueled population growth.[12]

Whereas this scenario is somewhat simplified — cultural and biological evolution is never simple — the general principle of population growth helps fill in the last piece of the agricultural origins puzzle: that of the timing. It happened quickly in the last ten thousand years because that was when population pressure led to the need for sedentism and the opportunities for coevolution. It was no accident that agriculture first appeared in the Mideast, one of the ancient world's population centers. It took a few thousand more years before agriculture sprang up in the Americas, the last continents to be colonized.

Stealing from the Rich to Give to the Poor

Agriculture hardly sounds "insidious," for it was one of humanity's greatest discoveries, allowing human populations the grandeur they have known for thousands of years. But not only was the human way of life changed forever — so was the face of the earth. Land was cleared for crops and grazing herds.

Wild plants and animals were pushed aside for their domesticated cousins. And the sixth mass extinction began in earnest, quietly and gradually overshadowing the earlier hunting and foraging effects on earth's biodiversity.

"Ecology" is a complex science that can have many meanings, depending on who is using the term. I like to think of "ecology" in the sense popularized by the famed German scientist Ernst Haeckel, who described it as "the economy of nature."[13] Ecologists often state that if ecology is an economy, then its currency is energy. Even in archaeology, "optimal foraging" theory focuses on modes of energy capture by communities that hunted and gathered their food and wood. And energy capture is at the heart of the success and impact of agriculture.

Most living things on planet Earth acquire their energy directly or indirectly from the sun. Without the ability of plants to assimilate energy through photosynthesis, there could be no animal life. Animals get their solar energy from the plant products they eat, or by eating other animals that have eaten plants. In the coevolution of plants and animals, many plants chanced upon a means of packaging the energy in their seeds, thereby attracting animals who in turn were nothing more than mobile seed-dispersal mechanisms for the plants. That packaging became more efficient, at least for the plant-human interaction, with domesticated plants.

By planting crops, humans were concentrating energy capture in a localized area. As foragers, they had to travel great distances throughout the year to harvest energy. Moreover, this energy had to be shared with birds, squirrels, bears, and all living things. Even the bits that fell to the ground were not wasted, for they fed a host of microorganisms (which in turn nourished the soil for more plants to grow). Many human foragers were able to store foods to ensure that the energy, once harvested, was preserved for their own consumption, and that was certainly true of the early agriculturists as well. But agriculturists could also protect their energy from other living organisms even before harvest — at least to a degree, for every farmer knows of the constant battles against crows who eat their grain or wolves who steal their sheep. But, minor losses aside, it was and is energy capture that makes agriculture work.

Sometimes it is possible to let other animals do the energy capture for us. Transhumance is a form of pastoralism in which people follow herds of wild animals, such as practiced by the Laplanders who follow reindeer in the northern reaches of Europe. Reindeer can eat lichens, which have never been a big component of the human diet. They transform the lichens' assimilated solar energy into their bodies, and the Laplanders then consume the blood, milk, and occasionally meat. What was in it for the reindeer? Protection from other carnivores, for one. The other attraction reindeer had to humans was a little more unsavory to the modern human mind. Like other cattle and other bovines, reindeer like saltlicks. The yellow snow left behind by humans provided just that. They were attracted to human urine.

Farm animals conduct the same services as reindeer — they transform energy from plants that are unpalatable or unfit for human consumption. Goats are particularly good at this, and thus they were among the earliest domesticates. When I worked at the fossil site of Taung, on the southeastern margin of the Kalahari Desert, I marveled at the ability of goats to eat leaves off the thorny acacia trees and find nutrients in just about anything. Indeed, a rusted-out old car along the dusty dirt road leading to the fossil site often had goats nibbling at it. Every year when I returned to Taung for a field season of excavation, the old car body was smaller.

As evidenced by the goats, animals assimilate other nutrients as well as energy into a form that is both richly energetic and palatable to people. From the standpoint of the human consumer, meat is the most efficient food source. Those who are vegetarian know that one has to eat a wide variety of plant foods to get the full complement of the vitamins and complete proteins with the amino acids that are necessary for healthy living. With animals as food, much of what we need is packed together as meat and easily digested.

But from an ecological perspective, in which we are concerned with energy transfer, meat consumption is inefficient. Less energy is lost along the way if we get it directly from the plants — the ecological component that Charles Elton termed "producers" in his classic energy pyramid (fig. 4.3). There is a net loss of energy involved in the growth and development of a farm animal. Sometimes we utilize a portion of that energy by having the domesticates work for us as draft animals, but for the most part it is lost to us. This is the reason that truly carnivorous animals are less numerous than those who consume plant matter — too much energy is lost going up the pyramid to support many "top feeders."

As human populations grew, they became more efficient at concentrating energy production for their own needs. For example, some learned to fell forests for agricultural land. A forest is much better at assimilating energy than a field of corn, but not at producing concentrated foods for human consumption. One reason that a forest is more productive is because of the biodiversity it supports. Tall trees gather most of the sunlight, but plants in other levels of the forest are adapted to produce with less sunlight. The ecological

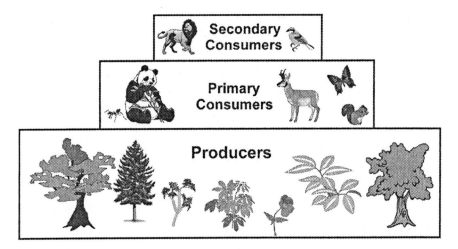

FIGURE 4.3 Ecologist Charles Elton envisioned a pyramid of life. Energy, the currency of ecology, is most abundant at the base of the pyramid, and is lost going through the food chain to the top. This limits the numbers of carnivores and other secondary consumers that can exist on earth.

interactions of many plants and animals maximize the assimilation of energy from the sun and nutrients from the soil.

Tropical agriculturists who use a form of production known as "swidden agriculture" utilize the principles of biodiversity to increase their crop yields. Typically they cut down a small swath of forest and burn the vegetation to enrich the soil. Then they plant many crops together, forming layers of growth that mimic the vertical levels of the rain forest, from the tall canopy on top to the smaller light-efficient plants below. What is lost in this system, however, is a replenishment of the nutrients in the soil. The soil can become depleted in one to three years, and then the farmers have to move on to another patch of forest and start the process over again. Meanwhile, the forest takes over the old agricultural plot, and does the natural work of making it productive again.

The same process is used, to less effect, in larger-scale agricultural practices. A single crop in a field quickly depletes the soil. Thus farmers must rotate crops and occasionally allow a field to lie fallow while the diverse species of nature unwittingly perform their services of soil production and enrichment. Alternatively, fields can be artificially replenished with fertilizers. Nature does the job more completely — it just takes more time.

Another way we increase agricultural productivity is with irrigation. Water is essential for life (hence the countless glasses of soda water I've consumed while writing this book). When water is diverted onto dry land, the land becomes more productive. But we don't tend to irrigate wild lands—we do it only for our own energy-assimilating domesticated crops. And if you divert water to one area, you have to take it away from another. It is essentially the same as felling a forest, in that we steal from the rich natural lands to give to the poor plants that form our food base.

Agriculture is the human way of taking over the powers of nature and concentrating its efforts on producing the resources we need. It is the most significant part of what is known as the "ecological transition."

The Ecological Transition

Cultural anthropologist John W. Bennett, from whom I was privileged to take a class during my graduate school days at Washington University, formally defined the ecological transition as "the progressive incorporation of Nature into human frames of purpose and action." [14] Professor Bennett's lectures were long stream-of-consciousness monologues that were sometimes hard to follow, but I seem to recall him also describing the transition as being from nature's control over humans to human control over nature. Obviously we don't have total control over nature, for as David Rindos put it: "The idea that we as a culture, a nation, or a species are in conscious control of our environment and thus of our destiny is one part truth, one part rhetoric, and two parts wishful thinking." [15] But certainly we have manipulated nature for our own purposes.

I like to think of the ecological transition as a long-term process. It started well over two million years ago with the shaping of the first stone tools. To be sure, our near-human ancestors probably *used* tools of various sorts before this time, but so do many other animals, most notably chimpanzees. But once early *Homo* employed the fracture properties of nature to fashion tools out of rock, a significant step had been taken. The stone tools became an extension of our own biology. Culture, or at least material culture, then shaped our own evolution in increasingly important ways.

The controlled use of fire was another huge advance in the ecological transition. Indeed, it was a giant leap more profound than Neil Armstrong's first step on the moon. We could harness energy for protection from predators, and cook food to make it safer, softer, and more palatable. Fire allowed early humans to venture into new environments without the slowly gained biological adaptations required of other living beings. Clothing and shelter did the same. We could shape our own environments and run circles around the leisurely pace of nature.

And eventually came agriculture, the most ambitious incorporation of nature into human frames of purpose and action. In the ecological terms of energy capture, Bennett states, "the transition concerns the tendency to seek ever-larger quantities of energy in order to satisfy the demands of human existence, comfort, and wealth." [16] Primary among the driving forces, as Bennett also notes, was the growth of the human population.

Of course, there was a cost. The ever widening human wedge sapped up energy that had supported other living beings. If you think about it, agriculture necessarily reduces biodiversity. A large field with a single crop has little diversity of either species or genes. Not only plants and animals but microbes in the soil are less diverse as well. [17] A farm field not only focuses large amounts of energy into one purpose — production for human consumption (either directly or indirectly through producing food for domesticated animals) — but also takes up space and nutrients. Those plants and animals that get wedged out do not necessarily go extinct, but their genetic biodiversity becomes diminished in smaller populations as the humans get more demanding of the land.

In the coevolutionary system of agricultural development, domesticated plants and animals evolved to become attractive to humans, not necessarily more nutritious. Some plants indeed were less nutritious than their wild ancestors or native counterparts. By planting fields of crops such as maize, people produced the bulk of food they needed to feed a growing population, but not the best type of food. That is one of the reasons human health declined. We see in the bones and teeth of early agriculturists the signs of malnutrition. Moreover, as noted earlier, people focused on eating a limited diversity of foods as opposed to the wide variety provided to the foragers by nature.

Domesticated animals were also limited in their capacity to cope. They became smaller as an adjustment to the crowding and limited diversity of feed provided by the farmers. Eventually animals were bred to focus their development on certain components such as milk production in cows and goats or egg production in chickens. This was done at the expense of other parts of their biology, hence the inability of many such animals to return to the wild.

The process of breeding plants and animals, like natural selection, limits variability. Yes, new forms arose, thus adding new species to the diversity of life on earth. But as the selection process continued, their genetic variability became less and less. The result is known as a monoculture, and the fields today are replete with monocultures of corn, wheat, and soybeans.

There is a problem with monocultures: they are quite vulnerable to attack by the ever evolving pathogens and pests. A fungal pathogen caused the potato blight in Ireland during the nineteenth century. Large expanses of potatoes with limited variability, imported as a crop from South America, were an easy target for the evolving fungus. It wreaked havoc (and is partially responsible for the McKee family being in the United States). Oddly enough, a relative of the fungal pathogen is now destroying oak trees in California.[18]

Unlike oaks, which rely on nature's services to reproduce and spread — such as squirrels burying acorns — most crops cannot regenerate themselves. If you leave a "healthy" cornfield to itself, it will not regrow itself in subsequent years. The same is true of the grass on our lawns, which cannot outcompete the weeds. It requires human intervention, and has spawned a large industry. We are the servicing agent in the environment of the crops.

To get around these problems there are a number of possible solutions. One has been increasing human intervention. Enriching the soil with fertilizer and controlling crop losses through the application of pesticide are thus forms of the ecological transition. Mind you, from an ecological perspective, this requires an imbalance of energy, for it takes more energy to produce and apply these artificial helpers of growth than we get in return. (Much of that energy comes from fossil fuels, but we'll save that story for later.) Moreover, the pesticides and fertilizers do not stay on the farm; they leach through the ground to the water table or erode into rivers and streams, thus disrupting the patterns and processes of natural ecosystems.

When Darwin's world meets the farmers' pesticides, microbes and insects can evolve quickly due to their sheer numbers and great diversity. Pests and pathogens ultimately evolve resistance to the chemicals, and an arms race proceeds. Problematically, they generate new defenses more quickly than the scientists trying to attack them can develop new chemical weapons. A more clever approach is to increase the genetic diversity of our crops. Scientists sometimes find the wild ancestors of today's crops, and introduce their long-lost genes into modern varieties. This renewed genetic diversity not only increases the resistance of crops to pests and pathogens, but can also be used to increase the nutritional value of the plants. (Inevitably, nutrients were lost in the initial domestication by unsuspecting agriculturists.) But what a pity if the human wedge has already led to the extinction of the wild ancestors!

Nevertheless, the ecological transition continues. Biotechnologists have found new ways to increase the genetic diversity of plants. They can select specific genes from unrelated plants, or other kinds of organisms, and basically insert them into the genomes of crops. The most common example is the insertion of an insect-killing protein from a bacterium into crops such as corn, thus reducing the need for chemical insecticides.[19] Such approaches have been used to augment a crop's natural pest resistance, increase yields, and improve the nutritional value of the food products. Yet such genetically modified or "GM" foods have provoked protests from a wide range of concerned citizens. Some refer to GM crops as "Frankenfoods," likening them to the monster created by Dr. Frankenstein in the classic novel by Mary Shelley. Although some of the protesters' concerns are valid, I look with dismay at many of the claims they make. GM foods are "not natural," they say. "They could harm the environment." But *no* food crops are "natural" in that sense: none have evolved in wild ecosystems. And the very nature of agriculture has always been harmful to the environment, as monocultures take the place of natural, diverse, and productive ecosystems.

All agricultural foods are "Frankenfoods," and have been for up to ten thousand years; GM foods are just the latest phase in the ecological transition. One wonders if the last of the human foragers protested the environmental disruption caused by their agriculturist counterparts. Wrote the late ecologist William Drury, Jr., "I think that the philosophical question of killing one species to favor another was made and accepted by those early agriculturalists who pulled up plants that inhibited the growth of their crops — they weeded the garden."[20] But what constitutes a weed?

The Mighty Weeds

Until the origin of agriculture, there was no such thing as a weed. A weed, by horticultural definition, is "a plant of no value . . . that tends to overgrow or choke out more desirable plants."[21] Many of the plants we now consider to be weeds predate the agricultural revolution; they were (and still are) just opportunists that invade disturbed soil. Indeed, a few such "weeds" helped initiate the agricultural revolution by "invading" the disturbed soil around human habitation sites (and then coevolving with the humans to eventually become "crops"). Moreover, by ecological definition, weeds are an important part of nature: as pioneer species, the initial invaders quite literally do the groundwork of establishing soils for the plants and trees that later follow a natural succession through various phases leading, in many regions, to the establishment of forests.

The problem with weeds is that they are better at capturing solar energy and soil nutrients than our highly selected monocultures of crops. They also reproduce and disperse quickly and effectively, as I see on my lawn every year. But from the point of view of earth's biodiversity, it is actually the crops that are the "unwanted" weeds. They deprive the land of more productive ecological systems and the biodiversity in them.

If we extend the definition of "weeds" to include animals, humans are the most effective weed ever known. As botanist John W. Bews put it in 1931: "Pioneer men, like pioneer plants, are most plastic in their reactions to the inorganic environment. . . . It is their business to conquer new habitats. They are ready to meet any emergency; they can fight their way through all the varied difficulties presented to them by Nature, but they fail to subordinate themselves to the community as a whole, when that becomes more complex."[22] As pioneer weeds, we capture an overwhelming amount of energy, and reproduce swiftly. Humans deprive the land of biodiversity, and would be largely "unwanted" by most other species (should they have the option for desire). However, sometimes when there are too many of us, nature automatically begins to weed its garden.

Population Pressure and Agricultural Collapse

It is somewhat ironic that population pressure, a key force in the origins of agriculture, has also occasionally led to the collapse of agricultural economies and cultures. Some would even say that this is happening today, although agricultural production seems to have kept pace with human population growth so far. That was not always the case, as archaeologists have seen from a number of places around the globe.

Today the Fertile Crescent is not particularly fertile. Part of the reason for the decline in productivity of the region was human intervention, through destructive agricultural practices that denuded the land: overgrazing, cutting wood for fuel, erosion, and so on.[23] But there is also a close correlation between climatic downturns around 4,200 years ago and the abandonment of the rain-fed lands of northern Mesopotamia.[24] Many northern people moved to southern Mesopotamia, where irrigation fed the fields. But that put greater stress on the land—irrigation leads to salinization of the soil. Year after year of pouring water over the land adds salts, and erodes nutrients.

Apparently, people also became stressed. The inhabitants of southern

Mesopotamia built a wall to exclude the hungry refugees of the north from entering their lands. Ultimately the society collapsed.

Peter deMenocal of Columbia University chronicled the changes many agricultural communities of the past endured in the face of climatic downturns.[25] Even the collapse of the great Mayan civilization appears to have been related to a combination of climate change and population pressure.[26] In each case, large areas became depopulated; only in some cases did nature eventually restore the land.

In some places the ravages of agriculture made it difficult for nature to restore the previous ecosystem. Salinization of the soil from irrigation is just one example. Even weeds have problems reestablishing nature's services to such lands. We can also change local climates: overgrazing of lands in the Sahel of Africa led to reduced rainfall and expansion of the desert.

Ohio was a region that saw a full recovery after an agricultural collapse. Atop a bluff downstream from my "office," the rock in the Olentangy River where I sit and think, are the remains of a fortress established about a thousand years ago. It attests to the large, successful populations of Native Americans that once dominated the region, growing corn, beans, and squash. But when Europeans first entered Ohio in the seventeenth century, they encountered a heavily wooded area with few inhabitants.[27] What had happened?

Part of the problem was the nasty side of humanity. As populations grew and productive lands became scarce, the protohistoric peoples of Ohio and the surrounding area turned their bows and arrows toward each other. Hence the palisades around many of the ancient villages. Indeed, the bluff-top fortress site 110 feet above the Olentangy River presents more than just an attractive vista: it also provides a commanding view for the defense of people's territory. In the protohistoric period, it appears that expansion of the Iroquois nation from the northeast led to raids on the Shawnee populations of Ohio. But that alone cannot explain the abandonment of this rich region.

Population pressure also brings disease. Just as tightly packed monocultures of crops are subject to evolving pathogens, so are people. Thus another part of the protohistoric Ohio population problem, once again evidenced by skeletal remains, was overreliance on corn. The resulting malnutrition not only had its own direct effects, but reduced human resistance to bacteria and viruses. Moreover, Europeans brought diseases to which native Americans had no immunity.[28] Millions died from tuberculosis and other diseases — that is truly how the West was "won." But the Ohio region had been largely depopulated before those devastating epidemics spread across the continent.

The final assault on early Ohio populations was famine. A drought one year would leave too little food to store. A flood in another year could wipe out the crops planted along fertile river banks. Although the native Ohioans were learning to deal with the unpredictability of nature, the ecological transition was not far enough along for them to have adequate control.

The archaeological record in Ohio and around the world reveals just what Thomas Malthus envisioned as the consequences of population growth: war, famine, and disease. The agricultural revolution was just one phase along the way. As much as we like to romanticize Native Americans living in perfect harmony with their environments, the archaeological record shows otherwise. From the overkill of megafauna in the early years to the subtle impacts of agriculture on wild species of plants and animals, as long as there were enough people, earth's biodiversity has had to give way to growing human populations. In some cases the expansion led to continuing extinctions of other species; in other cases there were setbacks, and humans had to give way to nature.

The perceived balance that Native Americans had with nature was due in part to lack of numbers: their populations had not yet grown to the extent of those in other parts of the world. Whereas native Americans grew to sufficient numbers to bring down a number of large animal species, the rest of the world had a head start at population growth and began spilling over into the Americas. What the Europeans found was a land that had not been transformed to the same degree as lands of the Old World.

Brown University anthropologist Shepard Krech put it this way: "The native people who molded North America were fully capable of transformative action in ecosystems they knew intimately, but in almost all instances their populations were too small to have made much of a difference. And when people few in number quickly became fewer from disease, the lands they had burned, cleared, and planted — lands transformed and exploited for purposes relating to agriculture, fuel, hunting, gathering, construction, and other ends — rested and recovered from whatever human pressures they had been under."[29] Given a few more years of population growth in the Americas, rather than decline from European diseases, the land may have looked very different upon the arrival of people from the Old World.

One irony of population pressure is that it led to the preservation of a large mammal species in at least one case. The bison that were once bountiful across the American plains were a favored target for Native American hunters. As the human population grew, the number of bison dwindled. But because population pressure can also lead to human aggression and war, there were often buffer zones on territorial boundaries between warring tribes. Thus, when Lewis and Clark took their famous expedition across the American West, they found that bison were plentiful only in the intertribal buffer zones, where the beasts were free to roam without human predators.[30] This of course changed when the expansion of European populations across the country led to wasteful slaughters of bison herds, pushing them to the brink of extinction.

Earth Out of Balance

When Europeans began expanding into Ohio, "one natural resource was both a hindrance and a necessity to the pioneer. Heavy forests of oak, maple, chestnut, hickory, walnut, elm, and other deciduous trees covered most of the state and constituted a serious obstacle to agriculture."[31] Whereas the trees provided wood for fuel and lumber for houses and furniture, much of the forests had to be burned down just to clear the land for planting crops. Ohio went from 95 percent woodland cover to a low of 12 percent in 1900. Now Ohioans must import lumber! Prairie land, though not the dominant ecosystem, was also converted to agricultural and residential land, cutting the amount of natural grasslands by as much as 75 percent. Wetlands were drained, such as the Great Black Swamp discussed in chapter 2. And with the forests, prairies, and swamps went the diverse species they supported, only to be replaced by monocultures of crops.

White-tailed deer were hunted and almost extirpated from the state. Despite this decline, as noted in chapter 1, the deer population rebounded in the twentieth century. It is worth looking into the causes more thoroughly, for this comeback illustrates well the complexities involved when humans and their agricultural practices transform the land. The near elimination of the deer's natural predators is but one factor leading to their recent successes. Those heavy forests cut down by the pioneers were not the favored habitat of the deer, but the "secondary" forests that have regrown in parts of the state are more open and conducive to deer browsing. Moreover, the deer like to eat corn. How do they get away with it? They have adapted by becoming more nocturnal. Now they are successful to the point where they must be culled in our parks, lest they destroy the "wild" vegetation. In short, we have totally altered the natural checks and balances among plant and animal populations.

Consequently, as a further phase of the ecological transition, we modern Ohioans have to take over the role of the predators.

In many regions of the world, in fact, people must now assume the role of ecosystem engineers. It is the next step needed to ensure that the further disruption and demise of our native wild species does not continue in the face of an ever expanding population.

Another option, of course, would be reducing our numbers, a solution nature has imposed a few times in the past. But today our population growth shows few signs of abating. Nor are we likely to abandon the land as population pressure grips us, as in the past, for there is nowhere else to go — and we are too good at manipulating our resources. Or are we? Can we sustain ourselves agriculturally, even with the effects of climatic changes and increasing population pressure? That depends on how clever we are and how we want to live.[32]

The world population is larger and wealthier than ever. We understand our nutritional needs better than did the early agriculturists. We are also more clever and innovative at harnessing and manipulating the ecological currency of energy. The green revolution is continuing, and seems to be defying the Malthusian vision of increased war, famine, and disease. It seems that we can rely on our innovative ways to continue such trends. My point is that there will be a cost. There always has been, as evidenced by the amount of ecological currency we already use. Self-styled auditor of the earth Stuart Pimm, a conservation biologist at Columbia University, estimates that by direct and indirect means, we already consume about 42 percent of the plant production from the earth's lands.[33] That doesn't leave much for other species to thrive — many cannot even survive. And the crisis is only getting more profound.

By the year 2050 there are expected to be about nine billion people on earth. They will all want food, water, and dignity. Taking into account the trends of success in increasing agricultural output, as well as the trends of population growth, a team led by David Tilman of the University of Minnesota looked at the possible impacts agriculture could have on our global environment.[34] We may be able to feed nine billion people, but the surface of the earth will have to change dramatically. Even with more-productive crops and more efficient farming practices, there must be an increase in the space utilized for agriculture. For example, the amount of land needing irrigation is expected to nearly double by 2050. This will require additional water from resources that are already being stressed in many areas. It will take water away from natural ecosystems, and can ultimately lead to salinization of the soil.

The water remaining available to natural ecosystems will be affected by runoff containing fertilizers, which can lead to an overenrichment of lakes and streams with extra nutrients, thus promoting ecological changes and loss of biodiversity in aquatic ecosystems.

In short, further increases in agricultural land must *necessarily* result in habitat destruction, just as has been the case since the origins of agriculture ten thousand years ago. Already human activity has transformed between 40 percent and 50 percent of the world's land surface for our own purposes.[35] Tilman and his colleagues show that if past trends continue, the amount of agricultural land needed by 2050 will be 18 percent more than at the turn of the millennium. An increase of this magnitude "would represent the worldwide loss of natural ecosystems larger than the United States."[36]

This is why the agricultural revolution was the genesis of a crisis: the diversity of life had to give way to less diverse plants and animals, in order to feed and clothe one expanding species. It is an insidious form of impact, because loss of diversity is difficult to see until it is too late. Noted Stanford University ecologist and evolutionist Paul Ehrlich commented on how humans have evolved to perceive imminent danger, such as a leopard jumping out from behind a tree, not to watch for long-term changes.[37] We relish our cornucopian agricultural successes, for we can eat and become wealthy in the here and now, but broader, long-term consequences tend to elude our consciousness. Meanwhile, other life-forms quietly and inconspicuously slip away.

Whereas some lands can recover from human impact — such as the forests of Ohio — recovery becomes less likely as the human wedge keeps growing and absorbing more energy and nutrients out of the global environment. And once a species goes extinct, it cannot rebound like the deer of Ohio. Extinction is forever. We'll see in the next chapter how human populations grow, and test if we can still observe today one simple correlation that seems evident from the past 1.8 million years and is clear over the past 10,000 years: human population growth is inextricably tied to the loss of species biodiversity.

5

C H A P T E R

Germs of Existence

NOT far from the Chinese fossil site of Zhoukoudian, where we have found the earliest evidence of our ancestors' spreading into temperate regions, is the modern city of Beijing. It is a sprawling city with an overwhelming number of people—well over fourteen million at the start of the millennium. If you have never been to Beijing, it is worth a visit. Not only can you see the cultural heritage of the Chinese people at places such as the Forbidden City and Tiananmen Square, but you can witness a more universal human heritage: the consequences of our population growth on the living world.

Once you get used to seeing masses of people everywhere, and learn how to dodge the hundreds of thousands of bicycles, you will probably be astounded to notice two things. The first is the sound of the city. Everywhere you go you hear an almost deafening *buzz*, loud enough to nearly drown out the din of constantly ringing bicycle bells. It comes from up in the trees, where the cicadas, far outnumbering the people, go about their business. The second thing to notice is the lack of birds in the air. I saw only one bird of any kind when I last visited, and it was in a cage being displayed for sale.[1] There weren't even pigeons on the statues. The overpopulation of the cicadas and the lack of birds are related phenomena.

In the 1950s the government of China foresaw a looming crisis. The human population was growing rapidly, and the ability of people to feed themselves was diminishing. Rather than try to halt the population growth—which was a government solution that came later—the leaders of China reached for another solution. "Man must conquer Nature" ran the slogan of Mao Zedong.[2] Sparrows were eating the people's grains, and were outcompeting people for the dwindling food supplies. They had to go. So birds were killed off by the citizens of China, their nests and eggs destroyed. Some of the birds were shot, but it was difficult to pick them off one by one. So, according to my Chinese hosts, people would go out every day and bang pots and pans to scare birds when they tried to land; eventually the birds died of exhaustion and starvation.

Of course birds eat more than just grain. They also eat insects. In the absence of predatory birds, the insect populations grew rapidly, before the Chinese realized — too late — the ecological devastation they had wrought by their policy of taking over nature. The insects became a threat to crops and destroyed the grain — a problem that continues today.

Now, despite the Chinese government's attempts to curb human population growth by forbidding couples to have more than one child, a program since abandoned, there are more and more people to feed. China is a large country, but much of its land is not arable — mountains and deserts are not conducive to growing food. So today the vast majority of arable land is taken up by crops or people. Hillsides are terraced, and crops are grown right up to the edges of the roads, and tended by hand. Little is left for natural species, including the giant panda and many more.

China is an example of how the human wedge has increased its impact through simple population growth. It is the ecological transition writ large. But why do populations still grow in the face of ever increasing pressure? It is for the same reason that the cicada populations grew — an excess of births over deaths.

Counting People

Throughout this book so far we've had to postulate growth rates of past populations to investigate whether or not human expansion could be related to the extinctions of prehistoric times. Whereas all these numbers have been hypothetical, they serve to show the possible outcomes of rates great and small. With time, even small rates of population growth could have had large consequences in terms of expanding the human wedge. For contemporary populations, the task of determining growth rates is considerably easier. Because most countries have census data today, we have a much better grasp of how many people are on the planet and how fast we are growing. The numbers are not perfect, for it is difficult to count everybody across such wide expanses and complex societies, but the margin of error is relatively small. Thus it is worth taking a close look at how those rates are determined and what impact different rates have on our future.

The key components of the growth are births (the fertility rate) and deaths (the mortality rate). For any individual country, immigration and emigration are additional factors; for example, in the United States, population growth is due more to immigration than fertility, although decreased mortality also plays a role.[3] But on a global scale, as the earth has no human immigrants or emigrants (excepting a few astronauts), we can concentrate on fertility and mortality rates.

In the past, and in many countries today, populations grew because the fertility rate was high. Like every other species, humans have evolved to reproduce as effectively as possible. That is what natural selection is all about: reproductive success with Malthus's "germs of existence." Fertility rates are usually expressed as the number of births per woman. They are derived from known rates of reproduction, which are used to calculate approximately how many children can be expected to be born of each woman in her reproductive years. Why focus on the woman? It should be obvious. But for those who like to think of us guys too, the "natality rate" is expressed as the number of births per year per thousand people.[4] So, for example, here in the United States in the year 2000, there were 2.1 births per female over her lifetime.[5] That is pretty close to the replacement rate, which would be one surviving offspring to replace the mother and one to replace the father.[6] By contrast, the fertility rate in my former home of South Africa was 2.5 that year, and in China it was 1.8. The natality rates can be found in table 5.1.

By the fertility rate alone, one would think that South Africa should be growing faster than China or the United States. But even without considering immigration and emigration, China is growing the fastest.[7] In other words, it has the greatest rate of "natural increase" (not that migrations are unnatural, mind you). How can a country with a fertility rate below replacement levels still grow? In part it can be attributed to the mortality rate, the number of deaths per thousand people per year. The relevant figures are also in table 5.1. Basically, relatively fewer people die in China. But before you run off to Beijing to find the magic elixir of longevity, we must consider the complexities involved in mortality rates. One is the age structure of populations.

Table 5.1 Population Characteristics of Three
Countries in the Year 2000

	Fertility Rate	Natality Rate	Mortality Rate	Growth Rate
United States	2.1	14	9	0.9%
South Africa	2.5	22	15	0.5%
China	1.8	16	7	0.9%

Source: The U.S. Census Bureau International Database

In the year 2000, China's population largely consisted of people in their reproductive years. That gives China great potential for population growth right now. Thus, whereas the Chinese fertility rate is the lowest of the three countries, its natality rate is greater than that of the United States. American baby boomers, the largest component group of the U.S. population, are slowing down their reproductive rate, which in part accounts for its lower growth rate right now. But watch out in a few years, when the "baby boomlets"—the sons and daughters of the baby boomers—are fully into their reproductive years. Depending on their life decisions, there is a potential for babies galore!

Population age structure also affects the mortality rate. Because there aren't proportionally as many people in the older generation of China whose lives are nearing their ends, the mortality rate is low. By comparison, the United States has an aging population, and proportionally more people are meeting their life expectancy. Clearly the mortality rate also depends on other factors, which vary in time and place: disease, war, famine, health care levels, and so on. Disease, and especially the AIDS epidemic, is a key reason that South Africa's mortality rate is so high. Yet despite AIDS, the South African population is growing—at least for now.[8]

Demographers gather all of this information to calculate a growth rate. For each country, they count the people, look at fertility rates and mortality rates, age structure, immigration, and emigration. Gathering the data is the hard part. Once it is all in a computer, it is relatively easy to combine all of the countries and come up with the worldwide yearly growth rate. As I write this, in the year 2001, that global rate is 1.25 percent. In other words, *if these patterns stay the same,* then in the next year our population of well over six billion people should increase by another 1.25 percent, translating into an addition of more than 76 million people. Remember, that is not the number of people born into this little world of ours, but the net gain of people.

Exponential Growth

When I lecture on population growth to my students at The Ohio State University, I have to provide them with a lot of data to get my points across. While writing this book, as in class, I've tried to limit how many numbers I display, for mathematics is not the forte of many people. In class I tell my students not to let all the numbers scare them; I just don't want them to shy away from coming back to class. But I should rephrase that comment for my students as well as for the readers of this book: let the numbers scare you. They are truly frightening in their magnitude and scope, and present us with one of the greatest dilemmas of the twenty-first century. So gird your loins, or whatever you do in preparation for a challenge, and read on—numbers and all.

In 1960, when I was just a toddler, the human population of the earth reached three billion. Now, one species—*Homo sapiens*—appeared on earth about 100,000 years ago. It had taken 100,000 years to get our numbers to three billion, though almost half of that count had been added since 1900. It only took 39 years to double to six billion, which we hit sometime in 1999. The near future will certainly see a population that adds people with ever increasing speed.

If you look at the graph showing global population growth in the past century (fig. 5.1), it is clear that we add more and more people every year. That is not because of an increase in the fertility rate or a decrease in the death rate; indeed, the growth rate has declined. It is because of the exponential nature of population growth. Exponential growth works by the same principle people use in establishing retirement accounts: they put in the same amount of money each year at the same interest rate, and expect to be millionaires by

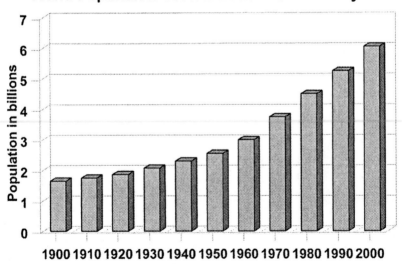

FIGURE 5.1 The twentieth century saw a remarkable increase in the number of people inhabiting the planet. Although the rate of growth has slowed, the exponential nature of growth ensures a continued climb in human population size.

the time they retire. More money makes more money as each year you add to the principal; likewise, more people produce more people, regardless of the growth rate. It is compounded interest. But those who incur excessive credit card debt or pay a mortgage know that exponential growth works in terms of debt as well. In our world, other species are the debtors, and many are rapidly headed toward bankruptcy.

Probably the best example I've ever read explaining exponential growth was written by "Weasel" of *The ZPG Reporter*.[9] He noted that in 1977, when Elvis Presley died, there were 48 professional Elvis impersonators. In 1996, there were 7,328. If this rate of growth were to continue, by the year 2012, one person in four on the face of the earth would be an Elvis impersonator. Profoundly odd, but it gets the point across. And the realities of population growth are even more frightening than a global glut of Elvis impersonators.

Right now, more than 200,000 people are added to the planet each day. Again, remember that we are talking about the net gain. It takes just over three days for the world population to grow by a number comparable to the entire population of Columbus, the largest city in Ohio. Since you may not have visited Columbus, I'll use a city with which more people are familiar: it takes about thirty-eight days for the net increase to reach a number equal to the population of New York City. I guess we need to build a new New York, and fast.

I have my students look at it another way. At the beginning of the academic quarter when I teach my Human Ecological Adaptations course, I give them the best estimate of the world population size on that day, as provided by the U.S. Census Bureau. Ten weeks later, when the course draws to a close, I provide them with an updated figure. Then I ask them to imagine that on day one we cleared out the entire state of Ohio. Every single person in Ohio's population of over 11.3 million had to leave from cities large and small, farms and countryside. As the world population grew by one person, we'd let one former resident of Ohio return home. This would happen at an immigration rate of about 146 people per minute. Before the time my lecture series was completed, everybody would be back home. By time the ten weeks was over, Ohio's population would have become greater than it was before we started!

I've heard many comments over the past few years from people who perceive an increase in the number of natural disasters in which hordes of people are killed. Earthquakes, monsoons, tornadoes, and so on *appear* to be on the rise. But they are no more frequent than they have been in the past. The truth is that our population has grown so much that more communities are living in disaster-prone areas. It leads to the perception that nature is keeping our population in check, but not so. If a natural disaster were to instantly wipe out five thousand people, the world population would quickly replace the victims in number (but not in character, of course). How quickly? Since the world population grows by about 146 people per minute, the numeric replacement would take just under thirty-five minutes. Even the grand forces of nature cannot hold back the germs of existence in such a large global population.

Even if the human population growth rate were seriously reduced, we would still grow at an alarming rate. For example, in the previous chapter I gave my calculation that if the growth rate had been constant since *Homo sapiens* first evolved, the figure would be 0.02033 percent—a fraction of today's growth rate of 1.25 percent. Applying that minuscule rate of growth to today's population, we would still add nearly a million people to the earth every year. That is the power of exponential growth, and the frightening nature of numbers.

Population growth rates are coming down, and *if* the decline continues we should eventually get to a point of zero growth. The United Nations predicts that should happen by 2200, with the world population at eleven billion people, nearly twice the number on earth today. But we are dealing with a very big if, as prognostication can be a dicey business. For example, subsequent analysis showed that there was an 85 percent chance that the world's population would stop growing by 2100, halting at a figure of about nine billion and then declining.[10] The projected declines in growth rate depend on a theory known as the "demographic transition," and we have to look closely at how this stabilization of our population is supposed to work.

The Demographic Transition

Like the agricultural revolution, the industrial revolution was also associated with population growth. Whereas I've argued that the principles of exponential growth may account for the "sudden" population increases some ten thousand years ago without any change in human reproductive habits, the story of Western countries between 1850 and 1930 certainly saw an increase in the population growth rate—the beginning phases of the "demographic transition."

The theory of the demographic transition actually started as an observation. Countries undergoing industrialization strove to use their technological

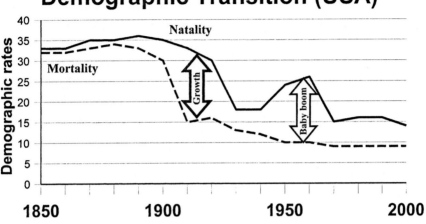

FIGURE 5.2 The transitions from high fertility (*top line*) and high mortality rates (*bottom line*) resulted in a spurt of population growth when mortality dropped first. This lasted until the fertility rate fell as well.

capabilities to make life better, and they did. Improved sanitation, advances in medicine, and enriched diets led to greater longevity. Not only did adults live longer, but infant mortality declined as well. Without any decrease in the fertility rate (and back then it was still not uncommon for a woman to have ten or twelve children), there was only one possible result: the population grew. And grew rapidly.

For some time there was a disparity between the high natality rate and the reduced mortality rate (fig. 5.2). Then something interesting began to happen: people started having fewer and fewer children. By sometime around 1930, the natality rate approached the mortality rate, the growth rate stabilized, and the transition was complete. In short, the observed demographic transition was one from high rates of births and deaths to lower rates of both. In the middle of the transition, the population boomed due to the disparity between birth and death rates.

Whereas it is clear why the mortality rate declined, the cause of the reduction in fertility is a matter of some speculation. The improvements in social and economic conditions seem to be at the heart of it. After all, there was less need of children for labor in the fields. And as people saw the riches of those around them, there was greater desire for wealth. Your money can't grow exponentially if you put it into children rather than in the bank.

One might suspect that electricity had something to do with it too. This was brought to my attention by an elderly gentleman in South Africa who was attending a lecture on the perils of population growth, given by Paul Ehrlich in Johannesburg. At the time, black Africans were still repressed by the apartheid regime, which allowed white Africans to hoard the wealth of the country. The gentleman who spoke was from one of the black villages where electricity was a rarity. In his best English, he commented on the high reproductive rate of "his people," stating that "with no electricity for lights and television, there is nothing else to do at night." Whereas television was not part of the industrial revolution, his point was well taken. With lights and wealth, there are other things to do at night.

Social changes also had an impact. Although it took time, the greater empowerment of women in society also helped to reduce the fertility rate. Rather than just get married, have kids (usually in that order back then), and spend one's day looking after them, women could make choices in their lives. Having fewer children was one of the options. Waiting until later in life to start a family was another, as women pursued educational and career opportunities. This does not mean that there was no pressure on them to conceive and look after the family; parents and in-laws would constantly ask if a "little one" was in the offing. But the choices women faced could be looked at in terms of the costs and benefits of children. Demographer John R. Weeks put it this way: "As fertility has gone down, more time has become available for women to pursue alternate lifestyles; and as the alternatives grow in number and attractiveness, the costs of having children have gone up." [11]

The demographic transition transpired for a variety of reasons, but the social and economic conditions did not strictly dictate fertility rates. Thus if you'll look back at figure 5.2, which is based on the pattern in the United States, there is a significant jolt to the fertility rate: the "baby boom" that happened after World War II as husbands and wives reunited and everybody was feeling optimistic and frisky. That rise lasted well into the 1950s. Moreover, the boom in births changed the age structure of the American population, thus increasing the fertility rate later when baby boomers started having their own families. It will continue to have an impact on natality rates in the future with the baby boomlets, sons and daughters of the baby boomers, coming of age. Nevertheless, overall fertility rates have declined enough to complete the transition.

The history of the demographic transition *may* be repeating itself. Many developing nations around the world are now apparently in the middle part of the transition. Improved conditions are leading to greater longevity and decreased infant mortality among formerly impoverished peoples. This is good — everybody on earth deserves a better life, and that is what many thousands of generations of *Homo sapiens* have striven for. On the other hand, the transition involves a lag in the decline of fertility rates, and thus most such populations are growing rapidly. As you may have surmised, my opinion is that this is bad when considered from a global perspective.

The theory is that the demographic transition will complete itself in contemporary developing countries, just as it did in what are now referred to as "developed countries" (though as I read the news each day, I often wonder about the sensibleness of that term). In some countries that appears to be the case — fertility rates are coming down. This is not always for the predicted reasons. For example, in Zimbabwe, just north of South Africa, fertility is declining along with wealth. The reasons may be complex.

Nevertheless, prognosticators see the demographic transition happening across the globe, and the world reaching a final population of between nine and eleven billion people. Frankly, I'm not so confident that the theory will prove true without proactive measures on the part of everybody. "The key determinant of the timing of the peak in population size is the assumed speed of fertility decline in the parts of the world that still have higher fertility."[12] Among the reasons for fertility declines are the proactive interventions of contraception and family-planning education. And even in most "developed" countries, the natural growth rate has not equaled zero. Knowing what we know about exponential growth even with small rates, I think there is cause for concern. That concern is for the other species on the planet that have not yet been squeezed out by our frightening numbers, but will be driven to extinction by eleven billion *or more* people. A human wedge of such proportions can only have negative consequences for biodiversity (fig. 5.3).

FIGURE 5.3 In Darwin's wedge model, the human species has become an increasingly important wedge in the living world as its population grows, forcing many other species into extinction.

What, Me Worry?

My favorite way to assess the impact of today's large population is to think in terms of the amount of space we take up. Not only agriculture, but housing, roads, waste treatment and disposal, commerce, industry, and recreation all take up a good deal of the surface of the earth. In the final chapter of *The Riddled Chain*, I posed this question: if we were to take six billion people and spread them evenly over the land surface of the earth, how much space would each individual have?[13] People have a hard time guessing the answer. Those who come from rural areas tend to think everybody would have a few square miles; those from cities usually guess much less.

Sometimes people answer by saying they heard that every person in the world could fit into the state of Texas. That is true, but each person would be rather uncomfortable in his or her own little plot of 10.74 by 10.74 meters — we'd all be within spitting distance of each other. Many people in the United States who cite this fact have heard it from Rush Limbaugh, nationally syndicated "conservative" radio talk show host.[14] Limbaugh, who quaintly labels anybody with an ecological conscience as an "environmentalist wacko," is trying to make the point that population growth is not a problem because there is so much space on earth. What he has really done is recast the "Netherlands fallacy" with an American state.[15] For years there have been those — the naysayers who are unconcerned about human population growth — who have pointed to the high population density of the Netherlands and claimed that since so many people can live there, then they can do so worldwide. But the people of the Netherlands must import their food and energy resources from elsewhere, and there is little room for wildlife.

So if we were to spread everybody back out evenly, how much space would they have? Each person would have just under 160 by 160 meters. I tell my students that that amount of space is less than the size of the Ohio State University football stadium. On a Saturday afternoon we can pack over 100,000 people into that stadium and still leave room for the players to have an open field. But then they all go home, and don't need to worry about making a living in such a small amount of space. Or do they?

Joel Cohen of Rockefeller University, in a very useful book titled *How Many People Can the Earth Support?* gathered a tremendous amount of data to try to answer that question.[16] We can use his analysis to fill up our football stadium with one person's needs. We'll assume that resources across the earth

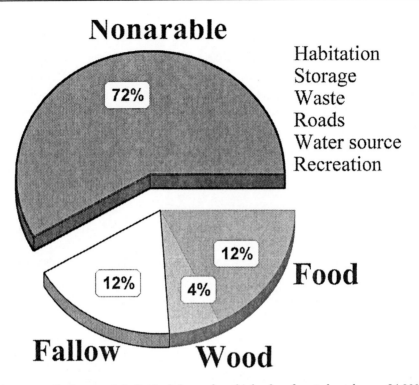

Nonarable

72%

Habitation
Storage
Waste
Roads
Water source
Recreation

12%

12%

4%

Food

Fallow **Wood**

FIGURE 5.4 Each person's individual share of earth's land surface today takes up 24,800 square meters. Most of the land is nonarable and cannot be used for agriculture. Once the remaining land is used for growing food and wood, only 12% is left fallow for nature to get on with its business.

are evenly distributed (which of course they are not) just to do this mental exercise. I've divided up the stadium in figure 5.4.

First of all, the biggest chunk of our stadium is useless for agriculture — 72 percent of the earth's surface area is nonarable land. That's okay for this exercise, because we can cleverly use that for other purposes: habitation, storage, waste, roads, water sources, and recreation. We could even share that space with plants and animals that do well in marginal environments: reindeer, scorpions, and so on.

Cohen estimates that the amount of land needed to grow food for one person is about three thousand square meters — about two-thirds the size of the playing field inside the stadium. This is assuming that we use fairly efficient agricultural practices, so we all need to bone up on our gardening skills.

We also need to focus on eating more plant foods (see Elton's energy pyramid, fig. 4.3) than meat, so that we can most efficiently harvest an optimal amount of energy. It is also necessary to have a high diversity of crops so that we can meet our nutritional needs, unlike the earliest agriculturists.

Energy is also a requirement. Those of us who live in highly "developed" parts of the world tend to forget that most people around the globe use wood for fuel. At my fossil sites in South Africa, my locally hired workers would often use their lunch break to cut firewood for their families—wood is a valuable commodity. Thus, if we want our share of energy from our individual plots, plus lumber for housing and furniture, then each person needs about a thousand square meters to grow trees. Hopefully some other life-forms can live there as well, but it is hardly the hundred-acre woods that Christopher Robin and his animal friends had to themselves.[17]

But if you have no need for wood, and would rather use metals for your construction or coal for your fuel, then you'll have to replace the wooded area with your share of the mines needed to extract the resources and your share of the power plant (which will consume a lot of your water resources). If you are clever, you can put up solar panels in the nonarable land, but you will still be capturing energy that could have been used for other living things.

Fortunately we have 12 percent of the land left to go fallow. But it is just a little more than we use for agriculture. Here is where the majority of species must live and conduct nature's services, although they get very few resources compared with each individual human.

The naysayers look blithely at such figures and say, "See? The earth can support six billion and probably more." They point to improvements in agricultural techniques that make the picture look downright rosy, for the average person. They know the other details, though: 840 million people are malnourished or hungry, 1.3 billion live in abject poverty, and 1.5 billion do not have access to safe drinking water and proper sanitation. This is dismissed as a cultural problem—with better distribution of resources and improved economic systems, we could overcome the shortages and everybody could live comfortably.

Naysayers also claim that we need not worry about actively curbing population growth, for the rate of growth is slowing of itself and eventually will halt. Even the United Nations sees the human population topping out at eleven billion, on the *theory* of the demographic transition.

For the sake of argument, let's say that they are correct—that we can feed eleven billion people, give each of them the dignity he or she deserves, and that the population will stop growing at that point. As much faith as I have in humanity—we are the most clever and adaptable species on the planet—I have my doubts that all this will be accomplished, even with eleven billion minds working to resolve our problems. But even under this optimistic scenario, there is one thing missing: where will the other forms of life go under the pressure of human expansion? Figure 5.5 shows how our individual plots of land would shrink with eleven billion people. Using the same estimates for assuring our needs, one slice is missing: the fallow land.

We need fallow land for nature to conduct all its tasks of regeneration. We need other species to conduct those services—decompose wastes, recycle carbon, build soil, produce the oxygen we breathe, and so on. Thus the important question is not how many people can the earth support? but how many people can the earth sustain? It is not just human beings that keep this planet alive, unless the ecological transition becomes complete and people manage nature in some brilliant, unforeseen way. But despite the best efforts of conservationists and concerned citizens around the world, our population

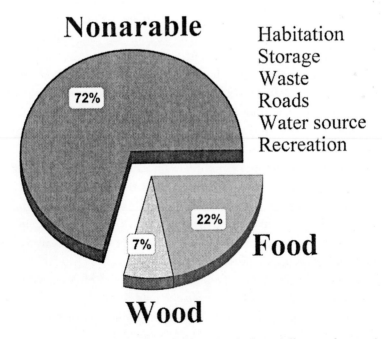

FIGURE 5.5 If we allow the human population to reach eleven billion, each person's individual share of land shrinks to only 13,535 square meters. The proportion of land needed for food and wood increases, leaving no productive lands fallow.

growth *is* wedging out other species—hence their impending bankruptcy. We'll have to look at some more frightening numbers to prove the point.

Gauging Extinction and Gouging Earth's Biodiversity

When we don't even know how many species there are on the planet, it is difficult to know how many are going extinct or to tie the losses to human population growth. Moreover, we know that extinction is normal. David Raup, esteemed paleontologist from the University of Chicago, noted: "For an evolutionary biologist to ignore extinction is probably as foolhardy as for a demographer to ignore mortality."[18] Extinction is necessary for Darwin's wedges to go through the progressions that have led to the assortment of biodiversity we have today, including the origination and evolution of our own species.

From the fossil record it is clear that most species that ever lived are now extinct; humans and every species surviving today are truly among the evolutionary elite. Noting that 99.9 percent of the species are gone, a statistician once quipped that "to a first approximation, *all* are extinct."[19] So in order to assess the magnitude of extinction today, we must assess whether or not the *rate* of extinction today is out of the ordinary.

Most scientists would agree that today's rate of extinction is significantly higher than in most of the past. Thus many people are astonished to learn that only a few more than a thousand species are recorded as having become extinct during recent times (since 1600). These include some well-known animals such as the dodo, a bizarre-looking flightless bird from Mauritius that went extinct by 1681. Another common example is the passenger pigeon, formerly the most populous bird in North America, which went extinct in the wild when the last one was shot here in Ohio in 1900. It later went completely extinct when Martha, the last one living in captivity, died at the Cincinnati Zoo in 1914. But these are just representatives of a host of lesser-known plants and animals of which most of us have never heard.

In South Africa, the quagga—a distinctive species of zebra with stripes only on the front half—was last seen in the wild in 1852. Before that, herds of up to one hundred quaggas graced parts of southern Africa, emitting the sounds that gave them their name. The last one died in 1875—at the Zoological Gardens in Berlin.[20] The quagga is one of nine mammal species of South Africa to have completely died out in historical times.[21] Nine doesn't seem like a lot, but we must put it in perspective. In my own research on South African fossils discussed earlier, I found an underlying pattern of about four or five mammal extinctions every 100,000 years. Here we have twice as many gone in just 400 years. That is an exceptionally high rate, especially since the evolutionary rate to replace such animals has not accelerated. Indeed, the evolution of new species must be slowing considerably, because their populations are getting smaller in the wake of human expansion across the country, resulting in a loss of the genetic biodiversity needed to generate new species. The inevitable consequence is a net loss of mammalian species biodiversity.

Keep in mind as well that my South African research deals only with mammals, because evidence of their existence can be most easily detected in the fossil record. There is no telling what the rates of extinction might have been for insects or plants or other living forms. But if the acceleration of extinction among them is comparable to that of the mammals, then the total number of historical extinctions in South Africa must be astounding. It is just difficult to put a precise number on it.

South Africa holds more than just fossil treasures. In the southwestern tip, near Cape Town, is the Cape Floristic region. In this relatively small region, no bigger than Ireland, are 8,500 plant species.[22] Most of those plants are endemic, meaning that they grow there and nowhere else. The predominant vegetation type is called fynbos, Afrikaans for "fine bush," which is a rich vegetational mix of grasses and shrubs with small and hard, leathery leaves, along with native flowers including ericas as well as the king proteas, South Africa's national flower. Fynbos plants thrive on the otherwise infertile soil of the region. Along with the unique vegetation are six endemic birds and nine endemic mammals. Nineteen of the reptiles and nineteen amphibians are endemic as well. At least that is the count today. In historic times, over three-quarters of the Cape Floristic region has been transformed by humans for agriculture and other purposes. We can't count what was lost, but we can count what would be lost if the remaining quarter were destroyed: a total of 5,735 species of plants and animals, found nowhere else, would go extinct.

Scientists from around the world are pooling their data and are starting to come to grips with global extinction rates. The First International Survey of Plant Diversity found numbers that give cause for serious concern. Out of *known* species of plants on earth, at least one of every eight is now threatened with extinction or nearly extinct. Goodness knows how many *unknown* species are at the brink. Much of this loss is due to deforestation in tropical areas, where land is being cleared for agriculture and other human demands. But it has not been limited to the tropics, for in many places such as the United States much of the land was cleared long ago, before there was significant scientific concern, and this led to countless "silent extinctions" — the loss of species that science has never known. But that has already happened. Many Americans, unaware of such past extinctions, think the problem occurs "somewhere else" and is therefore somebody else's problem. The current data should make them think again. The United States has probably the best-studied flora, and therefore the most "known" species of plants, but about 29 percent of those 16,000 plant species are at risk.[23] It is everybody's problem.

In terms of mammals and birds, about 11 percent of known mammals and birds are threatened with extinction worldwide. Should they go extinct, we will undoubtedly have the highest rate of extinction since the last mass extinction 65 million years ago. That, too, is compounded by the unknown — the silent extinctions that occur every day as we continue to clear the land.

To understand the magnitude of the silent extinctions, we must again extrapolate from the known to the unknown. We can use our estimates of earth's total biodiversity, and apply the same proportions of known extinction rates to those numbers. These estimates can then be honed by our knowledge of the patterns of biodiversity — the tropical rain forests are more diverse, and thus we can assume that the loss of species is greater there as the land is cleared. Using such knowledge we can get a good idea, but estimates of extinction rates will vary according to different underlying assumptions. Such estimates range from a loss of 0.6 percent of all species per decade up to 30 percent per decade.[24] Most estimates hover around a loss of 8 percent of all species per decade. Any one of those estimates constitutes a massive and rapid extinction event.

To imagine the impact of those numbers, we can again try to add perspective. Let's assume that the lowest estimate (0.6 percent) is correct. Using our biodiversity estimate of 12.5 million species on earth, that would mean a loss of about twenty-one species every day! It is not just giant pandas and gorillas but beetles and fish, yews and ferns. The demise of earth's biodiversity has clearly hit crisis proportions.

A Clear and Present Danger

Given the apparent long-term relationship between human population growth and biodiversity loss, it seems it would be easy to find a contemporary correlation between patterns of human demography and extinctions. But we live in a complex world, so the correlations are not so easy to find and test to scientific satisfaction. Such complexities quickly throw the naysayers into denial mode, even if common sense dictates otherwise. Yet persistence with difficult research has shown just what one might expect: human population growth today is closely tied to extinction risks.

In order to write this book on the relationship between human population growth and biodiversity loss, I had to convince myself that the connection could be verified statistically. Such research required amassing data from every country of the world on demography and biodiversity, so it was a daunting task. Moreover, previous attempts at testing population effects on biodiversity had met with only limited success, and found no connection between population growth rates and the threat of animal extinctions.[25] But I had to see the data for myself.

Fortunately, databases are more commonly being published on the Internet, and I happened to have an eager young student named C. David Fooce to collect and collate a variety of data. From each country, we collected information on population size and density, as well as fertility and growth rates, mostly from the U.S. Census Bureau's international database.[26] From the World Conservation Monitoring Centre Animals Database we got information on threatened mammal and bird species from each country. The WCMC categorizes threatened species by a complicated set of criteria, resulting in categories from most threatened to least threatened: critically endangered, endangered, and vulnerable.[27] We used these numbers to measure biodiversity loss: all such local populations of threatened species have relatively small numbers with little genetic biodiversity, and the imminent loss of some will also reduce species biodiversity.

It seemed obvious that the countries with higher population growth rates would have more threatened species. But like previous researchers, we found no such correlation. Unthwarted, and armed with a lot of numbers in the computer, we did some calculations to determine human population densities. After all, some country could have a high growth rate, but if few people were there to begin with, the impact would be small. But if people were already packed into a country, then the human wedge should be statistically visible, whether or not their population was growing.

In our initial test, we found just what we were looking for: a statistically significant correlation between human population densities and the numbers of threatened bird and mammal species. We rejoiced at the success of the mathematics, while commiserating about the implications. But upon inspecting the data closely, we found that the statistical correlations were highly biased by island countries. Islands tend to quickly acquire dense populations, and animal diversity is quickly squeezed out by the "weed mammal": humans. This has been dramatically illustrated in the past on islands such as Madagascar and New Zealand.

So it was back to the drawing board, or the computer circuit board, to make stricter tests of the relationships among continental countries. The correlations were still there, but the explanatory power of the statistics was initially somewhat lessened — until we looked harder. Indeed, inspecting the data carefully led to some interesting observations. Yes, some of the categories of threatened animals correlated with human population density, but they also correlated with average temperatures of the countries. We were also surprised to find that human densities and fertility rates were significantly correlated with the climatic variables! It seems that warm areas are conducive not only to mammalian diversity but to human reproductive rates as well. I don't want to hazard a guess as to why that is.

Thus our statistical tests had to consider environmental variables. We also had to look at the area encompassed by each country: larger countries have more animals, and thus have more potential for animals to be on the list of threatened species. Moreover, some countries have greater biodiversity than others, not only because of size but because of varying environments. So it turns out that human population densities also correlate with high biodiversity of birds and mammals.[28] Once we started taking all factors into account, using the best statistical techniques available, the simple correlations we had hoped to find began to reveal the complexities of the world we live in. So we would calculate and test, recalculate and retest. We brought in other colleagues to help us so we could think of all possible angles to draw useful information out of the data. No matter what we did, however, the correlations held up and sent the message loud and clear: there is an undeniable relationship between human population density and the number of animal species threatened.

More precisely, the correlation is this: as human population density increases, the relative number of threatened species per unit area tends to increase as well (taking into account temperature variations).[29] We were surprised to find that the population density in 1950 had a predictive value nearly equaling that of the density in the year 2000. In some ways this makes sense: there is a lag effect as humans encroach upon the land formerly inhabited by other species. Extinction can be a slow process from the perspective of the time each of us lives on earth, which is why it is so hard for some to see. But in terms of geological time, it is all happening in a blink of an eye.

As we were working on our analysis, Michael McKinney of the University of Tennessee published a research paper revealing the relationship between human population density and threat levels among mammal and bird species.[30] He had been thinking along pretty much the same lines as we were, and his numbers differed from ours only because he used different techniques. Our goal became to build upon such knowledge and find a way to take into consideration the most important factors to construct a mathematical model with some predictive power.

On the basis of our statistics, human population density in each country accounts for about 40 percent of the variability in the number of threatened animal species.[31] That may seem small, but it is an important chunk of accounting for biodiversity loss, given how many other factors might muddy the relationship. We live in a world of economically connected countries, after all, with their own histories of development, population growth, and varying capacities to harbor species biodiversity. For example, countries such as the United States or England have long had the "opportunity" to wedge out other species with cities and roads and agriculture. Countries such as Brazil now are growing rapidly, but the human consumption of land there also serves other growing populations through exports of lumber and cattle. Moreover, some countries are better understood than others — remember we are dealing only with known species in this analysis, whereas the unknown are probably under greater threat.

Still, if we could account for only 40 percent of the variability in threatened species across the world's countries, we had 60 percent to account for by other factors. Something more was needed to give our data better predictive value. So again we got to work, using our computers to try different predictive models and find those with the greatest explanatory power. We found one that still amazes us with its capabilities and simple elegance. Basically, a model using just human population density and species richness (number of species in a given area) is all that is needed to explain most of what we were after.[32] In other words, adding species richness accounted for many of the climatic and ecological differences among countries that help determine the potential for biodiversity threats. This simple model of just two variables can account for 88 percent of the variability in the relative frequency of threatened species across continental countries.

Most of our mathematical predictions for the number of threatened

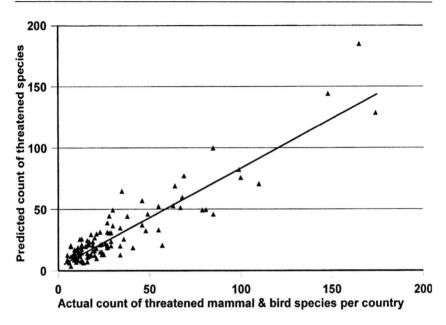

FIGURE 5.6 A simple mathematical model predicts the number of threatened species using data from each country on the relative number of mammal and bird species and the population density of humans. Most countries fall quite close to the prediction, indicated by the diagonal line, despite the complexities of ecology and culture.

species came pretty close to the current reality (fig. 5.6), usually within 1 or 2 percent. Brazil by far had the greatest miscalculation: we underestimated its number of threatened species by forty-five. That is still only a 2 percent error, given the high biodiversity. The error also makes good sense — as mentioned above, Brazil is destroying forests at an unprecedented rate to feed growing populations elsewhere, thus should display more threats than its native population would suggest. So whereas our model could not account for every nuance, it still possessed considerable mathematical and logical powers of prediction.

It was then time for us to turn our eyes to the future, and use the model to forecast what will happen to the number of threatened mammal and bird species if the human population continues to grow. For this we tapped data from the U.S. Census Bureau forecasts of each country's population size in 2020 and 2050. In went the numbers and out came the bad news. The average country would have a 7 percent increase in threatened species by 2020 and a whopping 14 percent increase by 2050.[33] That translates into an average of four species per country added to the threatened list in just fifty years.

The United States, even with its modest growth rate, is an exceptional country in this model—we can expect to imperil ten additional species by 2050 in association with population growth alone!

Whopping? you ask. Four species per country, even ten species, don't seem like enough to commiserate about.

But they are. With a perspective of evolutionary time, fifty years is just a brief moment for so many species across the globe to face their potential demise. And we are dealing just with known mammals and birds, not plants, insects, reptiles, or amphibians (whose populations are curiously shrinking across the planet). Keep in mind as well that our model is fairly conservative and considers human population growth *alone*, but not the increased exploitation of natural resources it would take to get all of us, in increasing numbers, up to a decent standard of living.

Furthermore, our predictive model does not take into account any cascade of environmental disruption, in which the loss of one keystone species leads to the demise of others, much in the same way that the loss of North American megafauna may have spurred the loss of other species in their ecosystems. If indeed there is a living domino effect, the true percentage of threatened species in the future may become even worse than our simple projections allow. The numbers of threatened species may increase exponentially, rather than in the conservative linear fashion we've modeled. We thus need a pension plan for nature, and like most such plans, the portfolio must be diversified.

Hot Spots and Chilling Prospects

In the global attempt to conserve earth's biodiversity, scientists have designated a number of places on earth known as "hot spots." These are "areas featuring exceptional concentrations of endemic species and experiencing exceptional loss of habitat." [34] Since the effort to designate hot spots was begun in 1989 under the leadership of Norman Myers of Oxford University, twenty-five hot spots have been identified around the globe (fig. 5.7). Most of them are in tropical areas where species biodiversity is known to be high, but others, such as the Cape Floristic region, are spread around other parts of the globe.

The idea behind the research endeavor is to focus conservation efforts on these parts of the globe to stem the mass extinction that characterizes our time. Because of the large degree of endemism, the twenty-five hot spots are

FIGURE 5.7 Twenty-five "hot spots" have been identified around the globe, represented here by the flames. These are areas of high species biodiversity imperiled by the human population.

home to 44 percent of all known plant species and 35 percent of known vertebrate animals.[35] The loss of that many species would be an environmental tragedy unparalleled in human history, and we have *already* put a pretty big dent in species biodiversity. The tragedy may already be under way: like the Cape Floristic region, most of these areas have already lost a significant part of their vegetative cover to human expansion. All in all, only 12 percent of the hot spots' original areas remain relatively undisturbed.

It is of some relief that a good portion of the hot spot lands, some 38 percent, are protected in parks and nature reserves. Yet laws and priorities change, so the effort to protect the lands must persist, especially in light of increased pressure from growing human populations. Richard Cincotta and Robert Engleman, of Population Action International, analyzed human population data from the hot spots and found that growth rates in these areas averaged 1.8 percent annually, as compared with the current world rate of 1.25 percent.[36] Only one of the twenty-five hot spots was experiencing a decline in population: the Caucasus of central Asia, a highly diverse temperate forest. Yet the Caucasus forest remains among the most densely populated hot spots, and has already lost 90 percent of its original extent.[37]

In some ways it should not be surprising that hot spots are densely populated and have higher than normal growth rates. After all, part of the definition of "hot spot" is that the species are in peril. Moreover, as my student and I found, the tropical regions are home to many of the most diverse areas, and these warm, wet regions seem to be conducive to high rates of human fertility, just as they are for plants and animals. The relationship is significant and important.

But is it fair to draw a direct link between population growth and biodiversity loss? It may be *what* humans do that matters the most, rather than how many of us there are. Yes, it is fair. Just as was the case with the Native American hunters who seem to have annihilated the large mammals ten thousand years ago, they could not have done it if there were not enough of them. To have a profound impact on biodiversity, we have to be populous as well as rapacious.

Whereas I applaud the hot spot strategy to conserve earth's rich and diverse biota, conservation alone will not be enough to save these lands in perpetuity. As long as populations grow, there will be pressure on the land. The evidence comes from many places around the globe, with a key example taking us back to the most populous country and the most celebrated species on the threatened list.

Poaching the Land

Ever since the World Wildlife Fund adopted the giant panda as their symbol, it has become an icon for conservation in general. But the giant panda is just an emissary from the animal world, representing hundreds of others that are threatened. It is rather like the Lorax of my favorite Dr. Seuss children's tale who "speaks for the trees." [38] The panda is what ecologists refer to as a "flagship species," for it draws public attention. One cannot expect to rally widespread public support to save a rare nematode (earthworm) or a threatened species of mosquito. But gorillas, giant pandas, and rhinos are visible, attractive, and threatened. So great attempts have been made to save the habitats of flagship species in order to preserve the lives of the other plants and animals that they represent, which are also under siege from human expansion.

But despite all of the attention to the giant panda, the unrelenting onslaught of population growth is a source of pressure that all the conservation in the world cannot stem. Such is clear at the Wolong Nature Reserve in China. The reserve was established in 1975, but the giant pandas had to share it with the humans who lived there. Those people reproduced, and did so at a higher rate than the surrounding population. In the twenty years since the reserve was established, the number of people increased by 70 percent, and the number of households more than doubled. [39] This was to the detriment of the panda habitat—it turns out that the best terrain for pandas is also a popular source of firewood. Once again, the ecological currency of energy plays an important role as humans subsume more and more.

The energy needs of an expanding population can also have long-distance effects on conservation areas. Such is the story of the Arctic National Wildlife Refuge in Alaska. One tends to think of it as just a bunch of caribou on the tundra, but there are many other mammals, birds, insects, and plants there. There are also a variety of habitats, including tundra, but also coastal lagoons, mountains, boreal forests, and more. It is one of the few pristine areas left on earth, and is certainly unrivaled by refuges and parks anywhere else in the United States.

The refuge was established in 1960, when there were just three billion people on earth, after efforts by a visionary grassroots group rallied public opinion. Twenty years later the refuge was expanded. Two more decades and the Arctic National Wildlife Refuge was threatened by the energy needs of a growing population — the world's population had doubled since the reserve was established. This time the energy source was not firewood, but the fossil fuels of ages gone by that lay deep underground. And the only way to get at that oil was to drill from the pristine surface. This became the center of a huge political controversy (which may still be continuing as you read this), for the oil rigs would have to be placed in the calving grounds of the porcupine caribou, whose herds are already under pressure from the natural rigors of life in the Arctic.

My aim is not to get into the politics, but to point out that as populations grow, people's priorities change. Terms like "reserve" or "refuge" may be fleeting. We need not poach the caribou or the giant pandas to put them under pressure — we just poach their land to the same effect, be it for fuel or agriculture or even recreation (as in our national parks). Moreover, not just genetic and species biodiversity decline, but ecosystem biodiversity gets threatened. People who live in a temperate region that has already been irreparably transformed can thus threaten an arctic region where they don't even live. That is one of the reasons our correlations between population density and threatened species were not stronger on a country by country basis — growing populations can have long-distance effects.

When I sit at my "office" and look at the woods around me, I like to think that they will be there forever. But that might not be the case in 2100 when (and if) the world has nearly twice as many people as it does now. People will need those resources, just as they did when the first European settlers entered

Ohio. For the woods to be preserved, we either need to instill an irreversible conservation ethic in future generations—which I see as unlikely—or stem the growth of the human population. The latter option is doable.

There are many more ways, some obvious and some more subtle, for human population growth to threaten reserves and the biodiversity they house. As a quick example, American energy needs have consequences after consumption: air pollution. The fouled air from coal-burning plants now threatens plant and animal life in the Great Smoky Mountains National Park. Exhaust from cars driving through the park exacerbates the problem. Whatever we do with the ecological currency of energy has consequences.

There is a clear link between human population growth and biodiversity loss, but that link is mediated by the way humans behave. It is thus worth looking at the ways humans affect other species, to see if changing our behavior would be sufficient to create a sustainable earth. Can we avert the crisis by cleaner, greener, earth-friendly living, or will population growth inevitably lead to greater losses of biodiversity? It turns out that Malthus's "great restrictive law" has many dimensions.

6

C H A P T E R

The Great Restrictive Law

WHENEVER I go back to South Africa for excavations and research at the fossil sites in the Makapansgat Valley, my thoughts range beyond the changes the world has seen over the past three million years. Much of note has happened since the bones of our distant ancestors became buried in the cave, but the present and future are of concern too. In today's fast-paced world, anyone who periodically visits this remote location can see significant environmental changes on an annual basis. The valley is rich in species biodiversity, from the insects that my entomologist colleagues were finding and recording to grasses and ferns that amazed my botanist brother. The baboons I so dearly love to watch are just one of a few primate species in the valley that persist from year to year. But as the human population pressures the valley, recent changes accelerate and become more obvious.

From the fossil site I can walk through the valley toward the forest edge. First I go past the crop fields and grazing land set aside for cattle. Occasionally I see the local troop of baboons saunter by, and I stare with fascination. The valley's farmers, however, are not so amused and frequently shoot baboons that are raiding their carefully tended crops. Like the birds in China, the baboons are competing for the grain that is meant for human consumption, and they have to go. Like rats in a city, South African baboons are treated as vermin.

Past the last farmhouse, I walk through an overgrown clearing. Reaching the edge of the forest, I stop, sit, and curse the "black jacks" that got stuck in my socks and were irritating my ankles. These are the seeds of a plant going by various names around the world, known locally as khakibos and elsewhere as beggar's-tick.[1] A single plant produces over three thousand seeds, and the black jacks ensure their dissemination with two sharp "teeth" that hook passersby. This well-evolved mechanism is as effective as it is a nuisance. As I pull sharp little black jacks out of my socks, one by one, I am unwittingly helping them take over the valley. Indeed, in the many years I've been visiting the valley, the khakibos have spread dramatically—on the socks of hikers, or the coats of baboons (who also spend hours picking them off each other). Far from helping spread biodiversity, the process is damaging diversity. It turns out that the plants are alien invaders of South Africa, and are outcompeting the local flora. They are native to tropical America and were accidentally introduced to South Africa by travelers (who probably had the seeds stuck in their socks). Pretty though their flowers may be in the fall, they are a weed and nuisance to the local plants as well as animals.

In the forest is a stream that cascades down small rocky waterfalls into natural pools that are wonderfully inviting for swimming. In years of drought, which are common in South Africa, the pools get low. This is partly due to the lack of rainfall, but also because more water is tapped out downstream or pumped out of the groundwater for the parched fields of grain. Usually a good year of rain replenishes the valley, but in recent years that has not been the case. It seems that the nearby town of Potgeitersrus was growing and the people needed more water, so they tapped into the groundwater of the Makapansgat Valley. This has the effect of sucking the valley dry, rather like pulling the plug in the bathtub and watching the water slowly drain. The sinking water table affects not only the plants and animals that live in this formerly pristine forest, but even dries up pools in nearby caves, where the translucent blind cave fish live in total darkness. Soon they too will be gone.

The years of drought that led to the draining of this valley have faraway causes. The droughts are most frequent during El Niño years, when temperatures rise in the tropical Pacific.[2] The underlying causes of El Niño, an irregularly appearing warm surface current in the southern Pacific, are not fully understood. Although it is a natural climatic phenomenon, its increasing intensity and frequency is thought by many to be related to atmospheric warming, which in turn appears to be caused by industrial emissions of greenhouse gases. All of the causes are far from the Makapansgat Valley, but such global climate changes trigger a complicated chain reaction that impacts the livelihoods of many a species who just need a bit of water.

If I'm lucky on one of my hikes, I can look up and see the valley's pair of black eagles soaring in circles above me. They are one of the few remaining pairs in South Africa, and rely on this patch of forest for protection and survival. But the forest itself is just a fragment of what used to exist in the country. Many species that lived there are now isolated in other patches far away. In some ways I am grateful for that—leopards used to be common in the valley. Many of them were hunted during the twentieth century until the valley had been cleared of this clever stalker. In a rare event, I witnessed a leopard take a baboon one night in the Makapansgat Valley, but usually they do not range this far from other natural enclaves. Nowadays the meat of the baboons, shot by farmers, nurtures nothing but flies and other tiny forms of life — not carnivorous cats.

Otherwise the forest looks good, and still manages to host a myriad of plants and animals, rare and common. At least for now. At the other end of the valley, near the main road, has sprung up a settlement of squatters. As the South African human population grows, and the government struggles to provide adequate housing and living facilities, people settle wherever they can. The problem, of course, is that the growing camp has no proper sanitation facilities, so the squatters pollute the land and the water.

The Makapansgat Valley is a microcosm illustrating how human population growth affects natural biodiversity the world over. Agriculture, irrigation, introduction of invasive species, habitat fragmentation, hunting, pollution, and global warming are key ways in which humans are wedging out other species. In this chapter we'll look at each of these and see how this pattern of human behavior has come to be so destructive, in ways both subtle and profound.

Slicing Up the Pie

Nature's great restrictive law has been in evidence since the beginning of agriculture. As we take up more land, water, and energy to produce our food, little is left for other species. At our current rate of agricultural expansion, it is estimated that by 2050 the worldwide amount of natural ecosystems converted to farms and pastures would total an area larger than the United States.[3] Because we take this land bit by bit across the globe, what is left for other species is small and widely dispersed patches of land.

If every human were given an equal share of the land — their stadium-size piece from our thought experiment — it would be difficult to get sufficient resources for sustenance and sustainability without agriculture. A forager would have to range beyond the boundaries to continue living. But this is in effect what has happened to many other species, for their resources are not evenly distributed either, and often they must travel far to get them. Yes, we set aside lands for them to live in, parks and reserves, but for many wide-ranging species this is not enough. Their habitats have been fragmented, and they don't have agriculture or any other means to focus ecological energy on producing the resources *they* need.

Habitat loss is clearly detrimental to other species, for they have no place to live. That is why changes in land use are seen as the primary driver of biodiversity disruption by an international team that looked at scenarios for the year 2100.[4] And there is an obvious link to human population growth — more people need more land and its resources, unless we drastically alter our behavior.

The problem of habitat loss is compounded by *habitat fragmentation* — dividing up species' ranges into patches, interspersed with environments that are less hospitable or even dangerous. This happens when roads are built, logging tracts are cut, agricultural lands expand, suburbs sprawl, rivers are dammed, and so on — the list is endless. Lifeways that have evolved over millions of years get cramped in tight spaces, and sustaining a wild population becomes difficult if not impossible.

Habitat fragmentation affects different species in different ways, but the net effect is a loss of biodiversity.[5] Some animals with more general rather than specific needs may survive the disruption to their habitat by altering their food sources and traversing the hostile territory between their preferred habitats. The squirrels that live just about everywhere around my home are a perfect example. Baboons are another good example of an adaptable species in the face of habitat fragmentation — they can cross roads, raid crops, and generally succeed if they are not shot first. The macaques of Asia, a monkey cousin of the baboon, indeed thrive in the disrupted environments and have pretty well become suburbanized! Their status in some places as a sacred monkey enhances their viability.

But for most species this is not the case, including one of our own cousins. The third most closely related genus to humans, the orangutans, are wide-ranging animals that have been subjected to habitat loss as well as habitat fragmentation. Orangutans, the "wild men" barely surviving on the islands of Borneo and Sumatra, were once more widespread across Indonesia and mainland China. Now they are becoming more and more isolated in fragmented patches as logging continues.

Orangutans don't travel much in a given day, but over the course of time they forage across vast expanses to find the scarce and unpredictable fruits they rely on. Like early foraging humans in many ways, they are quite clever at finding their food. They use their learned knowledge of the forest to track potential food resources, and can deduce the location of food by watching the movements of other animals. But their odd social structure is what really cramps their ability to survive in fragmented habitats. The males are quite solitary most of the time, excepting "long erotic treetop trysts" with the females when they breed.[6] The territories of the males overlap, but the whole habitat must allow suitable space for all the males to range and find an accepting partner for the tryst. Habitat fragmentation thus hinders mating. Even if the males locate willing mates, the orangutans' slow reproduction rate exacerbates their diminished potential for resilience. Their plight becomes increasingly difficult as fragmentation also aggravates intraspecific conflict. Although aggressive encounters between males competing for a partner have always occurred in this species, albeit largely vocal aggression, the frequency and violence of such encounters has increased as the available ranges become smaller and smaller.

The shouting will end when orangutans go extinct. But once again, they are just emissaries from the richly diverse forests of Indonesia. They speak for the trees, the plants, and the other animals that get lost in habitat fragmentation.

Many animals cannot find enough food, let alone mating partners, in a remnant patch of their formerly expansive habitat. This would be particularly true of large carnivores, the top feeders. At least early human foragers could turn to food sources other than meat if the hunting was not good, or if they drove their favorite mammoth species to extinction. But not so for wolves or jaguars. Even bears need their meat, although they survive on a variety of foods (indeed, I once saw a brown bear in Canada make a snack of a plastic catsup bottle). Such animals need to range quite far for prey, or else they might devour the entirety of a needed food source in a tightly enclosed area. That is why people in southern Ohio now find black bears in their backyards. Once extirpated in Ohio, the bears are roaming back to lands formerly theirs. They had to push beyond the bounds of the fragmented parks and natural lands that held them.

A further consequence of habitat fragmentation is that populations of species get packed into tighter quarters, and thus can more easily spread disease among each other. This is particularly noticeable among migratory birds who crowd onto the ever shrinking wetlands, spreading avian cholera and botulism. Even the mingling of birds at bird feeders has been shown to spread disease.[7] Now I'm not sure what to do with my feeder—I just wanted to help.

It turns out that the more specialized the species, the more vulnerable it is to habitat fragmentation. Darren Bender and his colleagues from the Ottawa-Carleton Institute wanted to know how habitat fragmentation has affected species populations. It is expensive and time-consuming to do ecological research, so they did what is known as a meta-analysis: an analysis of ecological analyses.[8] From twenty-five different studies, they were able to detect revealing statistical trends related to habitat fragmentation. For species that lived within a particular habitat, fragmentation led to a population decline that was greater than expected (as compared with what would be expected from simple loss of habitat area in more expansive territories). Like the orangutan, many species did not fare well in cramped quarters. Some species did well: the generalist species and some of the edge species, for the total amount of edge space around a habitat often increased with fragmentation. But edge areas tend to harbor fewer species, so there is usually a net loss of biodiversity.

The complexity of ecological systems is well illustrated by the meta-analysis. But the thriving of some edge species also jibes with a more subtle trend characterizing forest fragmentation. The edges of fragmented forest patches tend to recede due to the invasion of opportunistic weeds, such as plants and humans. When a road is cut or a swath of trees is cut down for new houses, the forest edge is rather abrupt; there is no transitional zone (called an ecotone) between habitats. In other words, there is no buffer to the forest interior. Thus secondary growth of less diverse weedy vegetation takes over at the disturbed edge, and the forest edge recedes.[9] This was clearly happening to the remnant forest at Makapansgat, where the khakibos were gradually working their way into the forest. The risk of fire exacerbates the situation, either from the fires themselves encroaching on the forest, or from the invasion of fire-resistant plants foreign to the forest habitat.

Whereas saving small patches of land is a noble endeavor, it is thus not a sufficiently promising strategy for maintaining biodiversity, particularly for species who need lots of space. More propitious results would come by keeping the human population from spreading into those lands in the first place.

Thinner Slices

When I was a Boy Scout, I loved to go camping at my scoutmaster's "farm." Once a productive farm, the land had become a mixture of woods and overgrown fields that he had wisely purchased and left for nature to take its course, just the opposite of what usually happens. We helped nature along, and planted rows and rows of pine tree saplings, which I am pleased to say have now grown dramatically into a woods. But there was an older tract of wooded land, and I was perplexed in my boyhood by how the scoutmaster had us clear out fallen trees. I distinctly remember asking, more out of fatigue than curiosity, why we should have to move out this wood when the death and falling of a tree is a natural process. The logs should just rot there, create good soil, habitats for bugs, and so on.

My scoutmaster patiently explained to me that, yes, natural processes usually take care of these things, but since it was such a small plot of land, we had to lend nature a helping hand. He wanted to avoid leaving around fuel for a devastating forest fire that would wipe out the nature of our campgrounds. Ohio used to be able to afford forest fires, as it was a part of the natural cycle. Now, we must be careful to protect our fragmented bits of habitat.

Ever since I was a boy, I've ruminated over my scoutmaster's piece of logic. On the one hand, I am sure that he was right. Now that our expanding human population has relegated the natural world to small patches, we must look after these bits of land. It is just a continuation of the ecological transition. On the other hand, have we learned enough about ecology to really manage those lands?

The case of a rare ground squirrel species in Idaho tells us something about management complexities.[10] These squirrels hibernate in the winter, and depend on fat-rich seeds of the native grasses to get them through the cold months. Natural, lightning-sparked fires benefit the squirrels, oddly enough, by burning off invasive plants that compete with the native grass species. Moreover, some of the ground squirrels' favorite seeds use heat to help them germinate. The system has worked for thousands of years.

But what do people do when the fire burns? They rush in to put it out, so as to preserve the remaining forests. It seems to make sense as a conservation measure. But it's not nice to fool mother nature. Our intervention is leading to the extinction of the ground squirrels. They starve without the growth of the grasses. Moreover, when a bedraggled population tries to migrate in search of mates, they are blocked by dense growths of pine trees, which thrive when the fire threat has been removed by well-meaning humans. The solution, of course, is to let the fires burn, but then we are at risk of losing the trees for the forest. The late Harvard ecologist William H. Drury, Jr., forcefully made the point that ecological communities are dynamic, and that change by natural processes is good for preserving a wider scope of biodiversity.[11] It just goes to show that conservation is not always the same as preservation. Sometimes we have to embrace fire as a natural part of ecosystems, and incorporate its controlled use into conservation management plans.[12]

The subtleties of habitat fragmentation go even deeper. Like the ground squirrels and orangutans who have difficulty finding mates, many species can barely keep viable populations going, and if they do, they have reduced genetic variability. Michael Rosenzweig, evolutionary ecology expert from the

University of Arizona, notes that the potential for adaptation and speciation — the regeneration of species biodiversity — becomes hampered by habitat fragmentation. Quite simply: "This loss of speciation will occur for two reasons: species with larger geographical ranges speciate faster; and loss of area drives up extinction rates, thus reducing the number of species available for speciation."[13] In other words, fragmentation is a feedback loop leading to a downward spiral of species biodiversity loss.

Will conservation keep us out of the woods in terms of biodiversity loss? No. It may appear that many threatened species can hang on, but Guy Cowlishaw of the Zoological Society of London has demonstrated that this is just an illusion. He was able to document a lag time between habitat loss and eventual extinction.[14] Extinction takes time, as breeding populations dwindle in their fragmented habitats. In particular Cowlishaw studied African forest primates. It is not just the endangered orangutans of Borneo and Sumatra but also many of the African primates that look set to follow the path of Miss Waldron's red colobus (see chapter 2).

Cowlishaw found an "extinction debt" for 30 percent of the primates in each country studied. In other words, populations were on the decline and could be expected to go extinct over the course of the next few decades — with or without conservation efforts to prevent hunting or further habitat loss. Hunting would accelerate the eventual extinctions, but none of the species was in danger from hunting alone. Habitat loss and fragmentation were the main culprits. The human wedge has come between interbreeding populations of once widespread primates, diminishing their chances to perpetuate future generations.

Michael Soulé captured the essence of the situation this way: "In other words, the extinction problem has little to do with the death rattle of the final actor. The curtain in the last act is but a punctuation mark — it is not interesting in itself. What biologists want to know about is the process of decline in range and numbers."[15] Whereas habitat loss and fragmentation are important contributors to the extinction process, more subtle effects of the human population can be equally devastating.

Another Fine Mess

Like orangutans, gorillas are among our closest living relatives. They have been under siege for some time due to habitat loss and fragmentation, as well as from poaching. Lands have been set aside for them and laws made to protect

FIGURE 6.1 Stand back when a gorilla sneezes. Humans and other primates share many diseases, and can pass them across species. (*Illustration by Timothy B. McKee*)

them, yet as gorilla populations struggle to hang on, wars in the African countries where they reside tend to circumvent the laws. Conservationists can no longer do their jobs, and gorillas are taken as "bush meat" by hungry soldiers. In times of peace, the strategy for their protection has been to encourage "ecotourism." Tourists pay a lot of money to go into the gorillas' native habitat and see them in the wild, and the local people get a slice of the takings to keep them happy and out of the gorillas' reserves.

It would be a wonderful thing to watch gorillas in the wild—I even love watching them in zoos. Just don't sneeze. Or cough. It is not that they will attack you because of your rude disturbance (although with their might, they could). It is because you will be spreading your disease. That is just what happened in Rwanda in 1988, and led to the gorillas sneezing, coughing, and dying (fig. 6.1). It seems they caught the humans' measles.[16]

Because the apes are so close to us in a genetic and evolutionary sense, they are susceptible to many of the same diseases. Yet until we got up close and personal with them, they had not encountered these diseases and had no natural immunity. So whereas the usual idea is to let the gorillas fend for themselves, veterinarians in Rwanda intervened and gave the remaining gorillas vaccinations. Ironically, now the world health community is struggling to contain and cure two deadly diseases that appear to have gone the other direction, from primates to people: HIV and Ebola.

Viruses and bacteria are pernicious hitchhikers, and they can evolve quickly. New variants spread rapidly, especially in such a dense and mobile population as modern humans have become. But because of the evolvability of these germs, they can easily take hold in other species. This spread is not limited to primates, but presents a particular conundrum for threatened species such as the orangutan. We can breed orangutans in zoos and prepare them for reintroduction to the wild, but there is a distinct danger that they will take our diseases back with them. Because of the threat of introducing potentially devastating disease, the only way to truly save orangutans in the wild is to ensure that they have sufficient habitat to proceed with enough romantic trysts and thereby sustain their population. The zoo animals cannot play a direct role, but only the indirect role of public education.

Biodiversity is *increasing* in at least one realm — that of human-borne diseases.[17] Viruses and bacteria have over six billion human breeding grounds potentially available for growth and transport. We combat their spread with our natural immunity, inoculations, and antibiotics. But given the way evolution works through natural selection, the antibiotics are killing off susceptible bacteria that would have competed with more virulent bacteria that are resistant to our antibiotic drugs — a lethal failure of policy.

Now you might just see this as a manifestation of Malthusian principles: perhaps these diseases will keep our population in check. So far they haven't, as we still reproduce more than fast enough to replace our losses. But the point here is that evolving diseases are not just our problem. The resistant strains can also tax the immune systems of other animals. For example, baboons living close to a tourist lodge, where they feed on human refuse, have more antibiotic-resistant bacteria in their systems than baboon troops that live in more isolated areas.[18] This is also true of animals less closely related to humans: rodents living in northwestern England have been reported to have resistant bacteria in their guts as well.[19]

Part of the problem is that antibiotics have not been used sparingly enough on humans. The other source of resistant bacteria, however, has been from the use of antibiotic additives in animal feed — not to keep the animals free from disease, but to encourage their growth. After all, there are a lot of people to feed with those livestock, and efficient development of farm animals benefits us all. But it also accelerates the evolution of new bacteria that may — or may not — affect the livelihoods of other, wild species. And the closer we get to them, or the more wild animals that visit our ever expanding farmlands, the greater the chances that such animals will catch something. Just like the gorillas who got the measles.

We carry with us more than just disease. As the human population expands it takes with it plants, pets, bugs, and all sorts of creatures. They too can spread their own diseases. But the problem goes well beyond one of introducing disease. It extends to the ecological disruption caused by the introduction of the new species themselves.

Invasion of the Aliens

It is bad enough for biodiversity prospects that we step in formerly pristine habitats with our own feet. But with seeds stuck to our socks, we bring weeds and new competitors into ecosystems that cannot handle them. Sometimes introductions of alien species are accidental; other times we have done it on purpose. But as our population grows and the global community becomes more mobile, we will inevitably take other species with us, virtually everywhere.

Intentional introductions would be almost amusing if they hadn't had such devastating consequences. Ettiene Leopold Trouvelot of Massachusetts had a great idea back in the 1860s. Or so it seemed. He brought in gypsy moth egg masses from France and cultured them on the vegetation in his backyard. The idea was to make silk, but the economic venture failed. And the entomological adventure was a disaster. Of course, some of his precious gypsy moths escaped, and the caterpillars born of the escapees began to defoliate the trees around his neighborhood.

Gypsy moths, as their name implies, get around quite well. They travel around by laying their egg masses on motor homes, camping trailers, and so on; thus we are the unsuspecting agent of their spread. Now there are gypsy moths here in Ohio, defoliating trees in neighborhoods and nature preserves. It seems that they cause more damage here because of the lack of natural enemies, such as birds and disease, that keep the numbers of other caterpillars in check. In other words, they are outcompeting the native species. For us to keep up with them, and prevent further damage, we have to spray our trees annually. And, like Trouvelot, we have not recouped the economic cost through silk production.

But it gets worse. As gypsy moths are on the rise, many native members of the silk moth family are on the decline, some threatened with extinction. Traditional explanations of DDT poisoning and habitat loss didn't seem sufficient to explain the decline. The culprit was *another* introduced species, a European fly that has a taste for silk moth caterpillars. The fly was brought over here and released in thirty U.S. states from 1906 through 1986.[20] The reason for the introduction: to control the spread of gypsy moths.

Another brilliant idea promoted by the U.S. Soil Conservation Service back in the 1930s was to use a plant known as multiflora rose as natural fencing for our farms. The idea was even supported by famed novelist, organic farmer, and conservationist Louis Bromfield, who lived in my home county and saw multiflora rose as a means to control soil erosion. Multiflora rose is an extremely prolific perennial shrub, native to East Asia, and quickly grows into impenetrable thickets. I never thought to curse Bromfield when, as a young boy playing in the woods, I got caught on the thorned branches of multiflora rose. Nor did I know it was an alien nuisance when I had to slash through the menace while doing archaeological surveys across the state. Multiflora rose seems to have succeeded very nicely indeed, thank you, and is now choking out our native vegetation of the Midwest.

Why do small, isolated populations of invaders suddenly spread with such success? After all, when discussing habitat fragmentation we noted that small populations are genetically impoverished with little chance to adapt. But biological invaders are usually fairly tenacious species with either effective means of dispersal, such as the black jacks stuck to our socks, or a capacity for quick growth, such as multiflora rose, which thrives on disturbed lands and forest edges. Moreover, because of the dispersal, they are venturing into new habitats rather than being isolated in an established one. What they find are weak competitors or prey that have evolved no resistance. The same principle applied when an invasive species known as *Homo sapiens* entered North America and found naive fauna to exploit, thus contributing to their rapid extinction. When the successful human hunters had established themselves, they in turn were devastated by the invasion from Europe of tuberculosis and smallpox germs, to which they had no evolved immunity. Not every species can be a successful invader, just a few—all the more reason that the net effect is a reduction of biodiversity. We see only the successes, and are blind to many of the losses.

Bunny Huggers and Honey Buggers

Sometimes environmentalists are put at odds against economic forces, and looked upon by the captains of politics and industry as mere "bunny huggers" who have a love for nature but little concern for how the human world works.

For some that may be true, but for most ecologists the threats to biodiversity are serious considerations spurring strong environmentalist stances. But perhaps the naysayers should take pause before they say much more. The ecology of invasive species can also have profound economic consequences.

About a third of the food you eat comes from crops pollinated by honeybees. Here in Ohio there are many orchards for which honeybees are the most manageable pollinators. But recently there haven't been many honeybees to pollinate them. Indeed, each year Ohioans have to rent honeybees, at great expense, to do the work. The honeybee populations were decimated by two mites that inadvertently made it to the United States from Europe and Asia. One mite literally chokes the bees to death by residing in the breathing tube and obstructing airflow. The other mite feeds on the blood of developing bees, leaving them deformed.

To add insult to injury, a beetle from South Africa made its way over here and has infested some beehives. It appears to have arrived with some South African soil. I don't think I brought it on one of my trips, but I easily could have had some dirt on excavation boots harboring a pest's egg just waiting to emerge and invade a new habitat. However it got here, the beetle, which is shielded from the bees' stings, crawls into the hive, eats the honey, and defecates. Not even the bees want to hang around in the fouled hive.

This was all very upsetting to me, for yet another species seemed to be imperiled by a biological invasion. Then I was surprised to learn that honeybees are not native to North America. There were brought here in the sixteenth century, and became vital to our agricultural system (which in turn feeds the growing population). With more land devoted to nonnative crops, and honeybees for competition, native bee populations declined—hence our reliance on honeybees. So this time, with the mites and the beetle damaging honeybee populations, the previous invaders became the new invaders' victims.

And what goes around often comes around. The mites that devastated American honeybees made it to South Africa. They were first spotted near Cape Town in 1997, and now have swept the country.[21] If they prove as devastatingly effective in their speed of dispersal as they did in the United States, then many of the native flower species—which heavily rely on bees for pollination—may be imperiled. This would include some of the endemic flowers in the Cape Floristic region, which were already imperiled by another invader: alien pine trees that spread from nearby plantations. But pine trees were brought in from Europe and North America for economic reasons, so lest I sound like a bunny hugger or a flower child, let me give one more example of the economic losses caused by alien invasions.

While humans are at war with each other, sometimes unseen enemies creep across borders. Such occurred during the chaos that accompanied the breakup of Yugoslavia. The "western corn rootworm" had slipped into Serbia, but the priorities of the people were on other battlefields.[22] The rootworm is not a worm at all, but a beetle whose larvae feed on the roots of corn and damage crops unless chemical pesticides are applied. It has now made its way as far as Italy, and is continuing to spread — at great expense.

Sleeping with the Enemy

Invasive species need not prey on others or outcompete them in order to reduce biodiversity. Sometimes it pays to keep your "enemies" close to you, especially if you can interbreed with them and influence them genetically. And in some cases this can lead to the ultimate extinction of your partner's species.

Normally different species are not supposed to be able to breed with each other. When they do, as classically illustrated by the donkey and the horse mating to produce a mule, their progeny are supposed to be infertile. But that is not always the case. Because evolution is a continuous process, species boundaries are often blurred. Thus lions and tigers can reproduce with each other — in zoos anyway — to produce "ligers" or "tigons." In the wild, comparable cross-matings can have a variety of effects.

In some ways the interbreeding of "separate" species can be a good thing. Darwin's finches provide one such example, as superbly described by Jonathan Weiner in *The Beak of the Finch*.[23] As conditions change on the Galápagos Islands from year to year, bird species can interbreed and quickly acquire beak traits suited to the predominant food sources. The interbreeding provides the genetic variability for natural selection to work upon. It is evolution in action, observed before our own eyes.

So what's the problem? With introduced species, or with species whose populations have been depleted by habitat loss or fragmentation, the evolutionary system can quickly go awry. For one thing, the genes of an introduced species can quickly swamp those of another with which it interbreeds. Such appears to be the case for the New Zealand gray duck. Humans may have hunted New Zealand's moas into extinction, but the gray duck may soon suffer the same fate because we introduced mallard ducks. More and more, the native duck populations take on the genetic and morphological features of the mallards, leading to losses of genetic biodiversity and eventually perhaps to an extinction of the entire native species — an extinction from within.[24]

For years environmentalists and loggers have slugged out in the courts their disagreements over the riches of the Pacific Northwest of the United States. At issue from the environmentalists' point of view is the preservation of old-growth forests that make up the home of the threatened spotted owl. And although the debate hits fever pitch sometimes, the spotted owls may be quietly breeding themselves out of existence anyway with the recently invaded barred owl. It seems that the two species have taken a liking to each other and can produce viable and fertile offspring. Thus the decline of the spotted owl, which is largely due to habitat loss, is exacerbated by an alien invader.

Interbreeding of species can also have very odd effects that confound our best intentions. Despite the species distinction between pet dogs (*Canis familiaris*) and wolves (*Canis lupus*), feral dogs can and will reproduce with wolves in the wild. That is not so surprising, since our dogs evolved from wolves in the selective process of domestication. Nobody knows what effect the interbreeding will have on wolf populations, but it does not bode well for the wolves, as many dogs had their cunning abilities selected out of them (in particular the ones I have had as pets). Nevertheless, it means that wolves could be delisted as endangered species and lose protection because they no longer satisfy the criteria of the Endangered Species Act. What contorted logic: we imperil them by allowing invasive species of our own making, and then cannot protect them under our own laws. To make matters worse in terms of rational decision making, wolf reintroductions to parks such as Yellowstone have pitted ecologists trying to reestablish natural ecosystems against nearby ranchers, trying to supply us our meat, who are justifiably afraid of what the wolves can do to their livestock.[25]

The interbreeding of species need not result in fertile offspring in order to have a damaging effect on one or both partners. In South Africa, the red hartebeest and the blesbok, magnificent antelope of different genera, will successfully mate, but their offspring are not fertile. Neither of them are invasive species, but they are thrust more closely together into smaller populations by habitat fragmentation. Because their offspring are infertile, they do not suffer the same genetic swamping as gray ducks or spotted owls. But they've wasted a lot of time and energy on creating and nurturing little bucks that won't carry on their lineage or keep the populations viable. Unfortunately, for these species, the buck may have stopped there.

The effects of interbreeding among species that have been brought or forced together go beyond the mere loss of genetic and species biodiversity. Across the United States, lakes and rivers are stocked with nonnative fish, much to the delight of those who fish for sport. Add to that the incidental release of alien bait fish by fishermen who know their art (I never got beyond using worms as bait). Not only can that lead to reduced variability through interbreeding, such as occurred among trout, but it leads to a "homogenization" of the aquatic systems across the country.[26] The fish of one river are more and more like those of another. What is lost is not only genetic and species diversity, but ecosystem diversity as well.

Show Me

Biological invasions disrupt natural ecosystems and reduce biodiversity, from genetic to species and ecosystem levels. That much is clear. Whereas problems such as habitat fragmentation are clearly related to population growth, how does the spread of invasive species relate to population growth? Despite our increasing numbers, if we became less mobile and were less intent on fiddling with nature through intentional introductions, then perhaps the problems would abate. But I doubt it.

Just thirty-five miles south of St. Louis, Missouri, downstream along the Mississippi River, is a quiet little town called Festus. There I once met an elderly gentleman who had lived in Festus for the entirety of his seventy-some years. We got talking about a wide variety of subjects, and at one point the city of St. Louis became a topic. "Never made it up there," he said. "I've seen it on the TV, but never made it up to the big city." And he lived only a half-hour drive away. No danger of him carrying a biological invader. Or is there? Ever since a Missouri senator uttered two words in Congress, "Show me" has become a catchphrase in the state. So we must show how population growth leads to biological invasions and why we can't all stay in one place like the man from Festus.

Festus lies amid a small farming community, so theoretically the people who live there have all the food and resources they need. Except energy. So they import electricity, which comes from imported coal. Up in St. Louis, over 2.5 million people live in a concentrated metro area where the only food grown is in backyard gardens—not much to live on. So St. Louis imports food. Not all foodstuffs can be grown in the temperate climate of Missouri, so food is transported in from other states, and from across the world. Thus from Festus or St. Louis, one can watch the barges go up and down the Mississippi River carrying food, energy, and who knows what else that may have attached itself for the ride.

In our thought experiment where everybody had their stadium-size plots of land, we assumed that the earth's resources were distributed evenly. Obviously they are not. We also assumed that habitation sites could be put in the nonarable land, and to an extent that is becoming more true. Thus cities spring up in desert areas where the people have to import water, energy, and food. In ecology, such imports are referred to as "unearned resources." It is not that people in these places do not work or deserve resources, but that the things that make our lives tick must be imported from elsewhere. Moreover, as cities such as St. Louis grow, the houses, shops, offices, and factories take over arable land, diminishing the possibility to produce "earned resources." Thus "unearned resources" become more common for cities or even densely populated countries such as the Netherlands or Japan.

Agriculture and energy production do not work without commerce, and commerce does not work without mobility. We have to keep on moving to support our large human population, and more movement becomes necessary to sustain a growing population. It's a small world after all, with so many people going so many places. And as we move about, more and more invasive species follow us, whether we want them to or not, be they as tiny as a bacterium or as large as a pet dog. Moreover, in the fragmented habitats caused by human expansion, interbreeding between invaders and closely related species becomes more likely as members of small populations look for suitable mates. With our large and growing population, biological invasions are inevitable, and the net effect is one of diminishing earth's biodiversity.

A Heated Controversy

"Don't like the weather? Just wait a while." So goes the phrase we use here in Ohio to describe our ever changing weather patterns. The quip usually refers to changes over the period of an hour or two. Now it means more. As each year goes by, we seem to have warmer summers with more record-breaking high temperatures, and the winters have been almost balmy at times, with less snow. Part of the reason, at least right here where I live, is the effect that our growing city has. Asphalt and buildings tend to absorb and hold the heat, whereas green plants would reflect it or utilize the sun's energy to help them grow. The rest of the changes are attributed to global warming.

Over the course of decades or millennia, climatic patterns change naturally. And year to year we can expect the unexpected, hence the phrase that started this section. But there is growing consensus in the scientific community that global warming is real and unusually rapid. Moreover, evidence is accumulating that the cause, or at least a significant *part* of the cause, is of human origin.[27]

One of the key pieces of evidence comes from the melting of mountain glaciers and polar ice.[28] This is a significant worry for those who live along the coasts, for as the ice melts the ocean levels rise and inundate low-lying coastal regions. The process is slow and hard to detect, so it is barely noticeable in the short run. Thus my parents, who live along a marsh on the coast of Georgia, have not taken me seriously when I tell them to sell their house soon. And coastal species that are already on the wane from the encroachment of human civilization have not yet been noticeably damaged by this additional threat.

In places such as the Arctic, where habitat loss and fragmentation are less severe, global warming is the chief threat to biodiversity.[29] Melting ice in the Arctic can be shown to have effects on species all the way through the food web.[30] At the base of the web are algae and plankton. With thinner ice, the algae are taking over, with less phytoplankton (tiny photosynthesizing organisms) to feed the zooplankton (protozoans and small crustaceans). The dead zooplankton usually rain down to the seafloor, feeding invertebrates at the bottom. Some of those are disappearing, including the clams, which in turn are food for the walruses. Fish also suffer, particularly the cod, who feed the seals. Fewer seals mean less food for the polar bears, as well as for those humans who also hunt seals.

As if the seals did not have it bad enough, if the ice breaks up early in the year due to global warming, then the pups of ringed seals get forced into the water before they are weaned and cannot fend for themselves. Early melting of the ice also affects the Peary caribou, who are already under pressure from Inuit peoples who hunt them. The caribou need solid ice to get from island to island, but now they are falling through the unstable ice. The same goes for bears who hunt seals on the ice; now they have to head for land before they have fed on enough seals to gain sufficient fat for the year.

The consequences of global warming also include increased droughts in some regions, such as those suffered in South Africa during El Niño, and floods in other places. That can be devastating to agriculture, but hey, we're humans, and we can adapt and redistribute our resources. Other species cannot. A significant flood in a small, fragmented habitat could wipe out the remaining populations of rare plant or animal species. Already, as their environments change, the species' isolation means that they cannot migrate to places where conditions are more appropriate for them (as they regularly did during climatic fluctuations of the past[31]). And drought can affect many marginal species such as the gelada baboon, one of the few open-land monkeys on the threatened list.[32]

Some life-forms will do well on a warmer earth, but their extra good fortune won't make you happy. Many infectious diseases will thrive under warmer conditions.[33] Most notable is malaria, which is expected to spread into more northern latitudes of Europe and North America.[34] Is this Malthus's predicted pestilence striking again? To a certain extent yes, but the rise of disease probably will not be enough to seriously stem our persistent population growth — it will just be enough to make us miserable. On the other hand, endangered species — such as the orangutans or gorillas noted earlier in this chapter — may be dealt a deathblow from disease. We just don't know.

Other life-forms, large and small, may also thrive with warmer conditions. After all, the decline of mammalian biodiversity we noted over the past two million years may in part be due to climatic cooling. When the tropics were more dominant on the earth, such as when the early caves of Makapansgat were filling up, life was richer and more diverse. There was more energy, the currency of ecology, to go around, and life abounded. Thus I often quip to my students that I am the only one I know who is *for* global warming. Just think of how biodiversity could thrive, as it did so long ago. The richly diverse flora and fauna of the tropics could spread farther from the equator. Temperate plants and animals could expand into the northern latitudes of Europe and North America. Even in the Arctic, as some species struggle, others could inhabit the land and increase the diversity; already it appears that the harp seal is on the increase.[35]

There is a catch, however. The threat to biodiversity is not so much that the planet is warming up, but the *fast rate* at which the changes are taking place. The current rapidity of global warming does not give many species time to adapt to the environmental changes it brings, making them more likely to go extinct. Sound familiar? The same argument was used by proponents of the hypothesis that quick warming after the last glacial phase led to the demise of North America's megafauna. Others argued that human hunting activities, augmented by the exponential growth of their population, were the cause of the massive extinctions. Or maybe one cause exacerbated the other.

Regardless of the academic debates about what happened ten thousand years ago, we can see both causes devastating biodiversity now: other living creatures are getting hit by the double whammy of rapid climatic change and human population growth. Especially fragmented populations cannot keep pace with the changing climate.[36] We can see it happening in real time. South Africa, already the country with the second-highest number of extinct plants due to human enterprises, faces even more extinctions with rising temperatures. By 2050, it is expected that summer temperatures in the interior of the country will rise by a few degrees Celsius, with lesser temperature effects — and less rainfall — along the Cape coast. Such changes will be devastating not only to the rare fynbos species of the Cape Floristic region, but also to savanna and woodland plant species.[37]

Will continued population growth accelerate global warming? The hypothesized causes of global warming are mainly our emissions of greenhouse gases, such as carbon dioxide, which trap solar heat in our atmosphere. Theoretically we could change our behavior, especially here in the United States, where our wasteful employment of energy makes us the world's worst culprit in greenhouse gas emissions. There are also attempts to find ways to remove carbon dioxide we add to the atmosphere, by capitalizing on the ways nature captures it in trees and other growing plants. The idea, basically, is to plant more trees if we burn more coal and oil. But we don't know if that can be accomplished. One thing is certain, all else being equal: more people require more energy, and it would be difficult to stem the tide of global warming by increasing the carbon capture of forests and prairies.

Even if we succeeded in altering our energy use to reduce our emissions of greenhouse gases, we'd have to change our agricultural priorities as well, particularly our taste for meat. The digestion process of ruminants such as cattle and sheep involves a process known as "enteric fermentation." One of the byproducts, coming out both ends of the animals in gaseous form, is methane — a greenhouse gas. Now you may not have considered cow flatulence to be a serious threat to the atmosphere, but countries such as Australia and Ireland have had to consider cutbacks on their livestock, their primary source of methane emissions.

On the other hand there may be an up side to the increase of one key greenhouse gas. Plants need carbon dioxide and can grow faster with enhanced levels of the gas, and that may boost crop yields. But the issues are complex. Experiments that raised carbon dioxide levels in pine forests have led to faster tree growth and greater production of cones and seeds, but the trees were smaller than usual when they reached maturity.[38] Moreover, some plants may do better than others in such an atmosphere. So, for example, if the pines responded well in a boreal forest, reproducing faster because of their increased seed production and fast growth, it could be at the expense of the birch and aspen trees (which would also then affect the species such as beaver that rely on such trees). So the up side may be down after all.

Thus the causes and consequences of global warming are complex. The ever rising tides along our coastal lands are but one of the problems, for both our economic well-being and ecology, that have alerted otherwise complacent people to the growing concern. But with melting ice caps, at least we'll have more water. Or will we?

7

C H A P T E R

Good to the Last Drop

WHERE there is water there is life. Water comprises about two-thirds of our human bodily composition, and accordingly is important in every facet of our daily survival. Most of our world is covered with water, hence the abundance of life on earth. Species are more diverse where there is more water and energy; energy alone cannot support life, hence the paucity of biodiversity in the world's deserts. The connection between water and all living organisms is why evidence for water on Mars is so important to the search for life on that planet. Yet despite the abundance of water here on earth, no vital component of life is more illustrative of the great restrictive law of how sparing nature can be with its resources.

Water was an important key to the origins of agriculture in the Fertile Crescent, and the lack of water was equally important to the fall of those early civilizations. It seems we still have not learned our lesson, for irrigation still leads to salinization of the soil or to erosion of the soil's nutrients. This leaves the land unfit for agriculture, and until a few pioneering species slowly conduct their valuable services of restoring the soil, it is unfit for high species biodiversity as well.

Agriculture also arose, independently, along the Yellow River about eight thousand years ago. Now this great river, flowing through the heart of China, is vital for the vastly overpopulated country. On its way to the sea, the river's waters are diverted into fields for agriculture and into the growing cities for consumption, sanitation, and industry. When I crossed the Yellow River in 1991, I was impressed by its majesty and size. But since 1985, each year the river has run dry for part of the year. In 1997, the river failed to reach the sea for 226 days. It had been sucked dry not by drought (although it had been a dry year), but by human consumption, leaving it barren for the aquatic species that used to thrive in its waters. Nature's forces had the last word. The following year heavy rains flooded the river, taking out farms, villages, and people. But the Chinese people returned to its banks as soon as they could. As long as the human population remains that large, they have to.

Most of the world's great cities are built along major waterways. Water is important for consumption and for transportation. One noticeable exception to the rule is Johannesburg, South Africa. There was no good reason for a city to spring up on the dry high veld, other than the gold that lay underneath the ground. Johannesburg was built on a river of gold, not one of water. Not only is water scarce there, but it takes a lot of energy to transport it and clean it, both before and after people use it. Thus water is a precious commodity in the city, and it was not uncommon when I lived there for us to undergo severe water restrictions. I was scolded for using running water overnight in my fossil preparation techniques, and could have been fined had I watered the grass. The great restrictive law had become a legal matter.

Now, despite the growth of cities along major waterways, more and more metropolitan areas experience water shortages. I never thought that would be the case here in Columbus, Ohio. We have two major rivers and three reservoirs to supply the city, and usually have abundant rains. I used to water my lawn with alacrity and let the tap run while I brushed my teeth or shaved. But water is no longer abundant for the taking, let alone the wasting. The Columbus metropolitan area, like so many urban centers across the world, is growing rapidly. Thus in a year that is slightly dry—which is when people tend to use *more* water—there are legal restrictions on water consumption. I had to let my grass go brown and subject my lawn to invasions by weeds that use the soil's sparse water more efficiently.

The solution most people propose to resolve the water crunch is this: build another dam so we can tap yet another reservoir for human consumption. I see an alternative solution as being more obvious: restrict the growth of the city. The populations of Columbus, Ohio, of the United States, and of the world are growing. People have to have someplace to go, whether to a city or along the banks of a river that is prone to floods. Building another dam can only accommodate them temporarily, but the long-term consequences are far-reaching.

Damning the Dam

I suppose that by now you can see where this discourse on water is leading. It should be obvious that as we take water for our own needs, it is diverted from other uses, such as the support of wildlife. Indeed, some national parks suffer when rivers are diverted or dams are built and disrupt the flow of water. Moreover, if we dam up the water, it has to flood another natural area. Nobody is going to dam up a river in the middle of a thriving city of homes and businesses, although dam projects usually do displace at least a few people. What is at risk? The last few areas where other species can live. At least that is true for land species; we'll get back to the fish and other aquatic creatures shortly.

The effect of dams on the plight of other species is illustrated well by the golden lion tamarin, a coastal monkey of Brazil. It is one of three subspecies of lion tamarins found only in remnant patches of the Atlantic forest region. Its habitat is inviting, and thus was one of the first parts of Brazil to be colonized over four hundred years ago. Given four hundred years of human population growth, the area is now densely inhabited. Moreover, the monkey's lowland habitat was easily cleared for agriculture. Many of the golden lion tamarins were displaced, others sent to zoos, others eaten by the local people.

Great efforts have been made to save the tamarins' genetic heritage in zoos, and to protect the species in the wild. In 1974, a reserve was established, mainly for the protection of this monkey. It was only nineteen square miles (5,000 hectares). But the nearby populace needed water, and part of the "reserve" was flooded by a dam. Today the golden lion tamarin remains on the critically endangered list. It is yet another example of how priorities change when the human population grows.

But it is just one species, some of you may say. Well, actually the golden lion tamarin is only one subspecies, but that is beside the point. What is lost with one type of animal are the others that depend on it. What is lost is an ecosystem. And that one subspecies is an anecdote, symptomatic of what is happening the world over on a daily basis—because the world's growing human population needs water.

Another counterpoint from some eco-optimists: "Well, at least those reservoirs provide a safe refuge for fish and other aquatic creatures." True to an extent, as long as we keep the water clean. But fish that live in reservoirs are different from those that thrived in the river valleys that got flooded by our dam projects. Moreover, what gets created is a large, uncovered expanse of water. This results in a larger surface area for evaporation, and we lose a lot of the water to the atmosphere. And just as with the evaporation of irrigation water, the salt precipitates lead to salinization of the lake bottom. This can be taken care of by the right combination of life-forms, but more often than not humans must intervene—all part of the ecological transition.

"Good-bye to a River"

Some of my college friends will remember one day in Cincinnati when I started quietly chuckling while looking at the Ohio River. Sitting atop a hill covered with buildings and roads, I saw the river as seeming so out of place. With all of our human accomplishments, the mighty Ohio River just kept flowing and there was nothing we could do about it. It was one part of nature we could not control. To be sure, we could build bridges over the impediment, and even use it for transport, but we could not alter its course to make it go where we wanted or keep it from flooding riverfront properties in the spring. I thought back to the comical winter when the Ohio was frozen enough for people to walk across it to the Kentucky side, but then got stranded as its persistent current broke down the ice. The river always had the last word and could wash away many a human folly. The river's power over us somehow amused my young mind.

Since that time I have come to understand more clearly how wrong I was about the river. In many respects it *is* under human control. Words from Don Henley's staggeringly beautiful and ominous song "Good-bye to a River" sum it up best: "Lakes and rivers, dams and locks, we put that river in a box. It was running wild, man must have control." Henley and the cowriters of the song captured the essence of the ecological transition. We do control the rivers and waterways for transportation, drinking water, waste disposal, and hydroelectric power—and thereby control the life in them.

Putting the rivers in a box has the same effect as habitat fragmentation, and indeed is an example of the effects. "More than 20 percent of all freshwater fish species are now threatened or endangered because dams and water withdrawals have destroyed the free-flowing river ecosystems where they thrive," wrote Peter Gleick of the Pacific Institute for Studies in Development, Environment, and Security.[1] Rivers are no longer running wild.

No fish species are more obviously affected and publicly discussed than the salmon in the northwestern United States, New England, and Europe. Although overfishing is part of the problem, the rest of the trouble for salmon comes from trying to get around the hurdles—sometimes literal ones—that the growing human population has put in their way. It seems that the juvenile salmon get blocked by the "boxes" of dams and reservoirs on their natural courses to the ocean, and others cannot make it back to their spawning waters—only a few can manage the journey. The spawning grounds require cool, clean water with plenty of gravel, but human activities ranging from logging to suburban sprawl have disrupted many of these habitats. The net result is that there are twenty-six kinds of salmon that are threatened with extinction, with further threats to their genetic biodiversity.[2]

The growing human population in California, and its incessant demand for energy—part of which comes from hydroelectric dams—has exacerbated the salmon's problems. In 2001, when an energy shortage hit Californians, the hydroelectric facilities in the nearby state of Washington had to release more water through their turbines to keep up with demand. Unfortunately, because it was a dry year, this left little water in the reservoirs to release in the spring, water that was necessary for juvenile salmon trying to make a run toward the sea.[3]

Of course salmon themselves provide us with energy directly as food. And they are quite tasty at that! They have been a dietary staple in the American Northwest since Native Americans began exploiting—and perhaps over-exploiting—the salmon thousands of years ago.[4] Today many of the food resources of the rivers and seas are pushed to the edge by excessive fishing.

One way to alleviate the problem of overfishing the salmon is to breed them in captivity. What results, of course, is a somewhat genetically homogenized salmon, selected for its ability to reproduce in captivity, just like other domesticated animals. That would be fine for us if it were not for one threat: if the captive fish get loose, then they interbreed with the wild salmon. The interbreeding would diminish the capacity of wild salmon to make their annual runs. Could such a thing happen? Well, yes, these things happen. In a human-dominated world in which we accidentally drop tons of oil into the ocean and cyanide into rivers, a few escaped fish is not out of the question. It has actually been more than just a few. Over 255,000 Atlantic salmon have escaped from farms in the North Pacific Ocean since 1980, some of which have been caught all along the Northwest Coast.[5] What effect their interbreeding will have on the wild salmon remains to be seen, for it may take many generations to sort out the genes.

Don't Drink the Water

Although the human population may act like one large sponge, rapidly absorbing the world's freshwater sources, we haven't done a very good job taking care of that natural resource. I live in a state that learned that lesson well in a distinctive way. The Cuyahoga River, running into Lake Erie near Cleveland, was so polluted that it caught fire in 1969 and burned uncontrollably for several days.[6] It is one of those ironically laughable images with tragic results. Quite obviously a burning river, let alone a polluted one, is not good for any living thing, including humans. It frightens me to think that I used to swim in Lake Erie back then, while on beach holidays with my family. It is a wonder the toxins pouring in from the rivers didn't poison our bodies forever.

Clearly pollution of the water or the air is not good for wildlife, so I need not belabor the point, save to quote the satirical songwriter Tom Lehrer: "Fish gotta swim and birds gotta fly, but they don't last long if they try." Fortunately people, governments, and industries learned the hard lesson of the Cuyahoga and other waterways, and we started cleaning up our acts. Now I've seen that the Cuyahoga can actually freeze during the winter, and Lake Erie is now clean enough that I let my own kids swim in it during the summer.

The problem is that it becomes more difficult to keep the water clean as the human population expands. This is illustrated by the beautiful lakes of the Boundary Waters Wilderness of the Minnesota-Ontario border. I was privileged in my youth to spend two weeks canoeing there in pristine wilderness. The water was so clean that we did not need to carry water canteens in our canoes — when we got thirsty we'd reach over and dip a cup into the lake. After portaging across short stretches of land, there would be another beautiful lake where we could canoe and watch the otters play. We were far enough from civilization to leave worldly concerns behind, and we were shocked upon returning to Ely, Minnesota, to find that the U.S. president had resigned and a new one had taken over. So you can figure out when I was there.

Now more people like to experience the natural feeling of getting away from it all (and no doubt some of them would hope to return to a new government). And traffic through the Boundary Waters Wilderness is starting to take a subtle toll.[7] No longer would I drink straight out of a lake, for years of people swimming in the lakes and doing other things in them as well has increased the lakes' bacteria populations — especially the kinds that pass through human digestive systems.

Of course fishing is a relaxing activity and a source of many a camper's meal. One subtle effect of this, however, is to remove biomass from the ecosystem, for decaying fish feed the lakes' natural processes. But the balance among kinds of the fish gets upset too. People throw back the little fish and cook up the big ones, thus altering the normal population structure of the fish schools. Moreover, people have a preference for walleye, lake trout, and northern pike — taking such fish preferentially throws off the natural balance of the community. This can lead to the demise of some species in the food chain, as has happened in so many aquatic ecosystems elsewhere.

Even something as wholesome as a campsite can have an impact on the ecosystem, especially as more and more people visit. The campsites lose their vegetative ground cover, allowing erosion of the soil into the lakes. Along with the soil go the remains of the campfires, concentrated in nutrients. Before the use of phosphate-based detergents was curtailed, phosphate additions to the lakes came from washing up after dinner. The process of building up nutrients in aquatic ecosystems is known as "eutrophication." Although that sounds good, and is often a natural process, it is kind of like eating too much candy. The human-accelerated buildup of nutrients throws a lake's ecosystem out of kilter, leading to the buildup of algae at the expense of other organisms that form the base of the food web.

Despite being far from civilization, the lakes are not free from industrial waste. It rains out of the sky, and acidifies the lakes. Sulfur dioxide, which makes sulfuric acid, comes from the burning of fossil fuels upwind. Rather than getting an "unearned resource," the Boundary Waters Wilderness gets unearned and undesirable waste. It either comes in the form of acid rain, or enters the lakes quickly during the spring as pollutants get released from the melting snow. Again it is the plankton at the base of the food web that get affected first, and later their loss can work its way up to kill off fish. Or the fish can be killed more directly. Even low levels of acidity can inhibit the spawning of fish such as walleye and lake trout, higher levels being toxic to more fish. So not even the fish in this remote area can "get away from it all."

Altered Courses

There are many ways we pollute the water and disrupt natural processes. For example, the runoff carrying soils and fertilizers from agricultural lands leads to the eutrophication of rivers and streams. But there are other, more subtle ways that just the very presence of numerous humans affects the water and the life in it.

Rivers are beautiful, and so people do more than just camp nearby. They like to live right next to rivers or put "scenic highways" along their banks. The problem is that in doing so, we have to clear many of the trees from the banks. Falling leaves from the trees feed microorganisms that in turn establish the base of the ecosystem (and clean the water for us). Moreover, trees provide shade for the river, and for particular plants and creatures. The shade helps to minimize temperature variation and ensure that life cycles evolved over millions of years are not disrupted. For example, numerous fish and amphibian breeding cycles are triggered by temperature changes throughout the seasons. For some amphibians, the gender of the new population is determined by temperature, so thermal imbalances one year affect the gender ratio and the breeding cycles in subsequent years.

Temperature is very important for many life-forms in the water, such as trout, but we don't usually think of one of the more insidious types of human disruption: thermal pollution. The most dramatic source of rising temperatures is industrial, but it is augmented by the tree loss that comes with urbanization and roads. What happens? With higher temperatures, there is a decrease in the solubility of oxygen, which is needed for the oxidation of biodegradable waste (from trees leaves or human waste products). To compound the problem, thermal pollution increases the rate of oxidation, so the little bit of oxygen remaining is depleted more quickly. This is not good for most species, as they need the oxygen to live. Some living forms benefit, however, such as undesirable algae that suck up oxygen more efficiently.

Some rivers have adjusted their ecosystems to thermal pollution, albeit with less diverse life. But one large source of thermal pollution is the water used for cooling the condensers of electric power generators. Electric companies have a habit of occasionally shutting down their generators for maintenance in the winter, which leads to a sudden decline in the water temperature. Some of the remaining life-forms of the river then die off from cold shock.

Mussels Muscling In

Humans use water for more than just consumption through drinking or irrigation. Natural waterways are also travel corridors for boats big and small. The importance of waterways almost led to a war between the states of Ohio and Michigan in 1835 when both coveted the Lake Erie port city of Toledo. Shipping is big business and an important part of human survival in this age, so key waterways are often worth fighting for.

Inevitably with the boats come hitchhikers, who find new habitats to exploit. This leads not only to homogenization of the world's water fauna, but also to alien invaders that can be quite disruptive to both ecosystems and economics, such as the zebra mussels that invaded Lake Erie and beyond.

Although seagoing barges and boats have been coming into Lake Erie for a long time, allowing invaders such as the sea lamprey to sneak in along with them, zebra mussels were first discovered in the Great Lakes in 1988. They are a fast-breeding mussel about the size of a fingernail, native to western Russia. After spreading across Europe over the past two hundred years, they made their way across the ocean in large ships, having been adventitiously picked up in ballast water collected at freshwater ports. The discharge of the ballast water spewed the mussels out into a new and receptive home in North America. They quickly spread during the next year, colonizing along every hard object they could find in Lake Erie and beyond. Such objects included the water intake structures of power plants and industries along the shore, which the zebra mussels clogged at great economic expense (over $5 billion by 2002, and counting[8]). They've worked their way to the other Great Lakes, with similar damaging effects.

It is curious that it took so long for the zebra mussels to find a home in the Great Lakes. Oddly enough, it may have been the cleaning up of Lake Erie that encouraged their successful bid for a takeover.[9] Now, as effective filter feeders, they have so enhanced the cleaning process that the water is no longer murky. What the mussels are eating is much of the phytoplankton and small zooplankton. So not only have they outcompeted the native mussels through effective and rapid breeding—just as multiflora rose does on land—but are disrupting the entire food chain by consuming so much at the base of the food web. Whereas the water now is lovely for sailing and swimming, it is a bit too clean for the aquatic animals that live there.

Of course natural waterways were not enough for effective shipping, so humans have constructed artificial channels to connect different watersheds. The Chicago Sanitary Shipping Canal is one such example, connecting Lake Michigan to the watershed that feeds the Mississippi River. By 1992, zebra mussels had happened upon such channels and traversed them effectively enough to work their way from the Great Lakes to the Mississippi, down as far as St. Louis, Missouri.

But the story of Lake Erie invasions does not end there: enter a fish known as the round goby. Gobies are also alien to the lake, and have quite an appetite for young zebra mussels—of which there are plenty. Like most successful aliens, gobies reproduce very quickly. And whereas they may gobble up many of the troublesome zebra mussels and allow native mussels to survive, they aggressively feed on other fish as well, particularly their eggs. It is impossible to say how the lake's ecosystem will respond next. As of now there are 145 alien species, ranging in size from bacteria and insects to plants and fish. The battle for Toledo is still being waged in the water by the relentless ways of nature, spurred on by our mobile population.

Ocean Harvests

When I was a boy, we often had Lake Erie perch fried up for dinner. To me it is not a fabulous-tasting fish, but it made a hearty meal. Apparently we were not the only family eating the yellow perch. It got overexploited, and perch populations plummeted in the early 1990s. Because restrictions were then placed on the fishing industry and sportfishers, perch populations are on the rebound in Lake Erie. Oddly enough, the perch also like to feed on the invasive round gobies, so they have an abundant food supply, at least for now.

When my students and I discuss the problems of feeding our growing human population, they often suggest that we need not worry about it too much, because there is an entire ocean of food for us that has largely been unexploited. They are usually surprised to learn that like the perch of Lake Erie, ocean fish and other food sources are already exploited to near maximum levels, and have been for some time.

Overfishing can have profound long-term effects threatening the viability of many species of fish, as well as those animals who eat the fish. A team led by Jeremy Jackson of the Scripps Institution of Oceanography combined paleontological and archaeological data with historical and ecological records to document the processes involved.[10] They could see evidence of impacts on local ecosystems whenever humans showed up, similar in many ways to those we have seen among mammals and birds. The ecological consequences of overfishing in the Americas were largely associated with European expansion (i.e., population growth), but smaller effects such as the decline of the sea otter along the Northwest Coast of the United States appeared to have causes thousands of years old. The Native Americans were not in harmony with aquatic nature.

Jackson and his colleagues make the point that overfishing has been more destructive to the ecosystems than pollution, invasive species, habitat destruction, and so on, because it is evident that the ecological disruption from overfishing preceded such modern maladies. It is not that such impacts are unimportant now, but rather that once again the long-term growth of the human population can take partial blame for the overexploitation of nature.

Large mammals who rely on fish are in competition with humans for food, and have now largely abandoned coastal regions worldwide. The effects of overexploitation also directly affect marine ecosystems. People tend to take the larger fish at the top of the food chain, so the initial effect is a decline in their populations, sometimes past the point of survivability. Once one species is gone, other similar predators become the target of overfishing. The big fish eat the little fish, so a decline in the number of aquatic predators leads to a burgeoning of the populations at lower trophic levels. These populations then become more dense and are more subject to outbreaks and the rapid spread of diseases.

The increase of low-level feeders such as oysters produced a lag effect of as much as two hundred years in Chesapeake Bay, showing the long-term consequences of disrupting ecosystems. The oyster populations of the bay, already substantial due to the reduction of their natural predators by fishing long ago, fed off the increased phytoplankton that grew from the increase in nutrients from agricultural runoff. The candy of eutrophication was great for the oysters, until they too were decimated by human harvesters. And as a final touch of disruption, the weakened ecosystems became prey to invasive species who found few competitors and open niches in the food chain.

Okay, so why not farm the oceans the way we do the land? True, we would have to take over a few natural ecosystems, but it would be more efficient than catching and disrupting wild populations. Or would it? It turns out that if "aquaculture" is practiced responsibly, it can compensate for the decline in fish available to be caught in the wild. Indeed, the contribution of farmed fish to the worldwide consumption of seafood doubled in the decade between 1987 and 1997. But ecosystem disruption is still a part of the game, just as in land-based agriculture.[11]

In ecology there is a strange term called "TANSTAAFL." It stands for the old adage "There ain't no such thing as a free lunch." It applies to our hopes for getting enough protein out of aquaculture to supply our substantial human population. Here's why: the farmed fish have to have something to eat too. Their food sources ultimately come from the phytoplankton in the ocean, and thus reduce the food available for wild coastal populations. Seaweed is brought in from elsewhere as well. Some farmed fish are fed smaller fish. Where do we get them? Largely from wild populations. And the Eltonian pyramid still applies: "Many intensive and semi-intensive aquaculture systems use 2–5 times more fish protein, in the form of fish meal, to feed the farmed species than is supplied by the farmed product."[12]

The habitat conversion is not inconsequential either. Aquaculture is practiced along coastlines and takes over many of the natural spawning grounds of fish that live farther out at sea. Coastal wetlands also serve to protect the coasts from erosion and floods, and aquaculture impairs their ability to do so. And of course, as noted earlier, the farmed fishes are often genetically different from the wild species or are not native to the waters of the farm sites, and have a tendency to escape on occasion.[13]

So why not go farther out to sea to harvest the riches of the waters that cover most of the earth? Well, it turns out not to be so rich out there. Ecologist Paul Colinvaux refers to the ocean as a "desert," for life is spread quite sparsely throughout it, excepting along the coastlines. As Colinvaux put it, "When people enter the sea in quest of food, they harvest the produce of a desert, which means they cannot get much, however large the desert may be."[14]

Sand and Soil

Because we have pushed the seas to their limits, or at least nearly so, the growing human population must rely on an expansion and intensification of land-based agriculture. Aside from the need to assimilate more and more natural lands where biodiversity flourishes, or used to, there are more subtle consequences of agriculture relevant to our discussion of water: erosion of the soil and evaporation of its moisture.

Soil is much more than just an accumulation of tiny fragments of earth. Soil is alive with microorganisms, worms, bugs, and roots, all cycling the nutrients and water. As Canadian scientist and author David Suzuki put it: "In this dark, teeming world, minute predators stalk their prey, tiny herbivores graze on algae, thousands of aquatic microorganisms throng a single drop of soil water, and fungi, bacteria and viruses play out their part on this invisible stage."[15] Soil has tremendous biodiversity within, breaking up waste into its component parts and fixing atmospheric nitrogen into usable forms; these activities are critical to the lives of species growing in and above the ground. We utilize those functions to grow our crops. But unlike the swidden agriculturists, who grow food in layers to mimic the plant diversity of the tropical forests, most farmers till the soil to lay it bare, and then sparsely plant a uniform crop. The first biodiversity to go is that of the microbes in the soil (which incidentally are also killed off by acid rain). Moreover, when the sun comes out it bakes the exposed soil, and when the rains come they wash the soil away.

The forests of Ohio once created a wealthy layer of soil, which inspired early settlers to give my home county the name Richland. It has served the agricultural community of the state well for more than a century. But each year as the crops are planted, more of the soil base erodes away, and very little accumulates to replenish it. Now soil is Ohio's largest export. It is not that we scoop it off the ground and ship it to other places in need. No, it just surreptitiously erodes away because there are no plants to hold it back. It goes into our rivers (adding to the eutrophication problems of aquatic ecosystems) and off toward the sea if not first clogged up in a water reservoir.

Farmers have learned to minimize erosion with sensible farming practices. Programs have been put in place to limit erosion, unfortunately including the introduction of multiflora rose. Nobody's perfect. Still the loss of good soil continues, as it has to, given the way we grow food. Between the erosion and salinization that comes from irrigation, many farmlands have had to be abandoned and forests felled for new agricultural plots.

The soil that remains also loses a lot of water and dries out, resulting in susceptibility to wind erosion. In times of drought the wind can be quite devastating, such as during the American "dust bowl" of the 1930s, when millions of tons of dirt became airborne and smothered the sky with huge dark clouds. Part of the problem was climatic, but a large portion of the blame could be laid at the hands of a growing American population that quickly stripped the land of its natural cover in order to grow wheat. Although we have learned our lesson in part, and now avoid such methods of dry-plains farming, the fact remains that no agricultural practice is as efficient as nature at maintaining the land.

We need not even till the land to affect the soil. Our cattle can overgraze the land and tromp across the remaining vegetation. In western Africa, south of the Sahara, is a region known as the Sahel. Droughts in this area are not unusual, but have been intensifying in their severity and length. One proposed cause is overgrazing by cattle.[16] Not only does it denude the land, but it alters vegetative cover that in turn impacts local weather patterns by reducing transpiration and allowing exposed soil to heat up. The result is an expansion of the nearby desert, or "desertification." Likewise in Sudan, the Sahara expanded south one hundred kilometers between 1958 and 1975, and has not receded since.

Desertification is a problem the world over, in China and North America as well as in Africa. Not only do we lose land formerly employed for agricultural purposes, but this forces us to expand farming into formerly wild territories so we can feed our growing population. The wildlife loses out even more than we do from desertification, for where there is *not* water there is *not* life. At best the diversity of life persists in diminished form, extracting the minute amount of water available.

Overgrazing that denudes the land was at the heart of a classic 1968 paper by Garrett Hardin of the University of California, Santa Barbara. He called it "The Tragedy of the Commons." What he meant by that title is best portrayed in Hardin's own words, though shortened here:

> Picture a pasture open to all. It is to be expected that each herdsman will try to keep as many cattle as possible on the commons. Such an arrangement may work reasonably satisfactorily for centuries because tribal wars, poaching, and disease keep the numbers of both man and beast well below the carrying capacity of the land. Finally, however, comes the day of reckoning, that is, the day when the long-desired goal of social stability becomes a reality. At this point the inherent logic of the commons remorselessly generates tragedy.
>
> As a rational being, each herdsman seeks to maximize his gain. Explicitly or implicitly, more or less consciously, he asks, "What is the utility to *me* of adding one more animal to my herd?" . . . The positive component is a function of the . . . proceeds from the sale of one more animal. The negative component is a function of the additional overgrazing created by one more animal . . . shared by all the herdsmen.
>
> Adding together the component partial utilities, the rational herdsman concludes that the only sensible course for him to pursue is to add another animal to his herd . . . but this is the conclusion reached by each and every rational herdsman sharing the commons. Therein is the tragedy. Each man is locked into a system that compels him to increase his herd without limit—in a world that is limited. . . . Freedom in the commons brings ruin to all.[17]

One might think that Hardin's point is that we should restrict the number of cattle by dividing up the commons among the herdsmen so that each will avoid the temptation to have too many animals overgrazing his own land. Indeed Hardin points out that we have done that by establishing private property as a rule for most of the world. But as our human population gets larger, an equal share of our private property gets smaller, and the end result is the same. In essence, humans are the overgrazers. Thus Hardin's real point, which we'll return to in the final chapter, is this: Can we continue the common freedom in the realm of *reproduction*, which brings more and more "grazers" to the human "herd"? Do we want the desert to keep expanding and water shortages to continue?

Let It Rain

The oceans not only have food for us to eat and natural ecosystems for other species, but they have another vital resource needed for life on land: water. Indeed they contain about 97 percent of it. The problem of course is that it is salt water, which is not usable for drinking or irrigation. Fresh water is the key to sustainability of life on land.

Ultimately most of our fresh water comes from the oceans. It evaporates from the ocean surface and rains down on the continents, feeding lakes and rivers and the ground where our crops grow. Of course we can capture fresh water directly from the oceans through artificial desalinization plants, but that takes space and energy. New technologies are being developed to make desalinization more efficient, but so far artificially desalinized water from the oceans accounts for just 0.2 percent of the fresh water we take for human purposes.[18] Nature does it better.

The problem with nature is that global weather systems and past geological events have not distributed the water evenly. Here in the Midwest of the United States we have it pretty good in terms of replenishing our water supplies through rain (although that is changing with global warming). The glacial phases of the past gouged out huge chunks of land and left behind their melting waters when they retreated ten thousand years ago. The Great Lakes now have 20 percent of the earth's fresh water. It is water that other people in less fortunate regions covet, and that we must share with other natural ecosystems. The question is this: as the human population grows, will there be enough fresh water to go around? Finding an answer is complicated, but the answer appears to be no.

Already we use about half of the earth's usable fresh water, most of which gets replenished by rain and melting snow.[19] Joel Cohen points out that it is difficult to calculate how much water there is and how much each person needs, but as the brave warrior of population statistics, he has made some calculations that at least give us a feel for the situation.[20]

Agriculture is the biggest user of freshwater supplies, so it is reasonable to focus on water and crops. Cohen estimated that in the rosiest of scenarios, with efficient agricultural production and consumption, and if *all* the renewable fresh water were used for agriculture alone, then the earth could support roughly 137.5 billion people. Hey, no problem for population growth, unless one starts thinking about the estimate. First of all, that would leave no water for direct consumption, sewage, industry, and so on. We need to *drink* water, especially in warmer regions. Sure, it is possible to get water from the crops we grow, much in the way baboons living in the desert go without drinking for days on end by eating foods high in liquid content.[21] But even the baboons drink copiously when they find a water supply. The second problem, from the perspective of this book, is that if humans use all the fresh water, then other living beings have none. But that is why Cohen calculated a variety of estimates.

If only 20 percent of the fresh water on earth were used for agriculture, which is somewhat more reasonable, then under the "best" scenario the earth's water could support 30.5 billion people. But Cohen's best-case scenarios are purposefully optimistic rather than realistic. As one considers different estimates of the variables involved, such as how much water is *conveniently* available or how efficient we are with its use, then the numbers start to drop. In Cohen's worst-case scenario, there is only enough water on earth to support 1.1 billion people. We passed that number shortly before the beginning of the twentieth century.

Indeed, the twentieth century saw a ninefold increase in the total amount of water humans drew from rivers and aquifers.[22] Meanwhile the population grew sixfold. Part of the disparity between six- and ninefold comes from increased need of water for irrigation rather than dry farming and for industry to support the growing population. Such a time perspective can be informative about our future plight. By 1995, a third of the human population was living under conditions of relative water scarcity, with about 450 million people in a situation of severe water stress.[23] So maybe Cohen's lower estimate was not so far off the mark. Projecting this into the future, given current trends of population growth (exacerbated by global warming), three billion people will be dependent on irrigated land by 2025 (as compared to two billion in 1985). Irrigation takes a lot of water, and the number of people under *severe* water stress rises to two billion. There is no telling how many plants and animals will suffer the same stresses, or worse.

The conclusion of a University of New Hampshire study was that "impending global-scale changes in population and economic development over the next 25 years will dictate the future relation between water supply and demand to a much greater degree than will changes in mean climate."[24] And given the uneven distribution of clean fresh water, the shortages suffered by two billion people will require radical solutions. Greater efficiency in using water is of course one solution. Moving water about by pipelines is another. Novel means of moving water helps as well, such as the large polyurethane bags that have been used to haul fresh water from mainland Greece to its islands to slake the thirst of an increasing number of demanding tourists. But we can't transport large amounts of water to the Boundary Waters Wilderness or spray it over suffering tropical forests. Nature has to do that, if we let it.

A famous quip often attributed to Mark Twain is apropos: "Whiskey's for drinking, water's for fighting about." The battle lines have been drawn by the sparing nature of the world's freshwater systems, and it is not just a battle for Toledo. In the Makapansgat Valley, the water taken by the nearby town of Potgeitersrus means that there is no water for irrigation of the farms, and the land has to lie fallow—and dry. The TANSTAAFL principle applies. It is not only we humans who lose yet another food source, but also the native plants and animals of the valley. They lose the battle. The complex chain of events that comes with human population growth leads to staggering effects, even more so than Twain's whiskey.

There is only one sustainable solution that I can think of: stop the growth of the world's human population.

8

C H A P T E R

Biodiversity in Action

PRIOR to the current wave of extinctions that is engulfing our planet, the living world has seen five mass extinctions. Despite the precarious imbalance that resulted, life on earth recovered and got back to previous levels of biodiversity, albeit slowly. In some ways this was good for us humans. For example, I'm just as glad that I don't have to run from a hungry *Tyrannosaurus rex*, although the image provides fodder for exciting cartoons and movies. Moreover, humans almost certainly would not have evolved had the demise of most dinosaurs not left the world open for the small mammals that eventually gave rise to our own ancestors.[1] Life on earth would be very different today, and we would not be here to take the blame for the demise of so many species.

Mass extinctions of the past bring up some vexing questions. Since they have happened before, are they not part of the natural order of this planet? After all, I've emphasized that the extinction of species and the generation of new ones is as normal as the life and death of individuals. Something is always left behind that seeds the next evolutionary era. Won't life recover its diversity through the evolutionary process even though it is temporarily depauperate of biodiversity? Aside from the discomfort of not knowing what position (if any) that humans would hold in a new world order, it seems reassuring that we cannot easily kill off a whole planet no matter how many of us there are.

Whereas hordes of early human hunters decimated the North American fauna thousands of years ago, both the Native Americans and the later European explorers found a rich land teeming with life. Today we still talk about the value of "pristine" wilderness, although no lands are truly beyond human impact. Such places just seem that way because they hold diverse species in a smoothly functioning ecosystem. So maybe our impact hasn't been that bad, and we can afford to lose a few more species here and there. The loss of species may sadden the bunny huggers, but the earth's ecosystems will march into a new era and take us along for the ride. Or will they?

In the opening chapter I made the following claim: conserving biodiversity is vital to the health of our planet, and consequently is vital to us. Throughout other chapters I have touched on the evidence and reasons behind this bold statement. Now it is time to pull the evidence together to show that there are practical reasons for conservation, and to demonstrate that our pension plan for nature must include a diverse portfolio. Of Charles Elton's reasons for conservation — religious, aesthetic and intellectual, and practical — only the practical reasons go beyond bunny hugging to logic necessary to convince an overpopulated world of people to slow down their reproductive habits.

We need biodiversity for a sustainable planet that includes human beings. Biodiversity is necessary at all levels — ecosystem, species, genetic, and everything in between. Without it we too can get dragged into extinction like the mighty *Tyrannosaurus rex*. But that is no way to solve our overpopulation problem. If I am sounding "alarmist," it is only because there is genuine cause for serious alarm. So let's look at the value of biodiversity at different levels. There is no time to lose.

Who in England Needs a Rain Forest?

Charles Darwin was enthralled by the tropical forests he saw during his famed voyage on the H.M.S. *Beagle.* Undeterred after being drenched by a rainstorm, he satiated his curiosity with the boundless myriads of Brazilian bugs, many of which he sent back to England. "I never experienced such intense delight," he wrote.[2] No doubt the rich biodiversity of the tropics played a role in helping Darwin formulate his views on the evolution of life. The tropics encapsulate the aesthetic and intellectual reasons for conservation. And there are practical reasons for keeping their ecosystems thriving.

Environmentalists spend a lot of time focusing the public's attention on tropical forests. Indeed tropical forests make up the majority of the world's "hot spots."[3] They are continually chopped down for agriculture, even though such activity has yielded only about 13 percent of the world's cropland, at the expense of a much higher proportion of species.[4] The value of tropical forests is not just the high species biodiversity that characterizes such parts of the world, but the vital services such ecosystems perform, which are necessary for preserving the way of life of people like us, living thousands of miles away. Saving the tropical forests is more than just a popular cause useful to politicians and rock stars seeking publicity, although it serves that function well. Hey, everybody has to have a cause, and this is an important one.

The immediate benefit of tropical forests is that they provide timber and fuel for those who live there. But forests do much more for the people around them. They create and maintain the soil that we are converting to agricultural land, and prevent the soil from eroding away. On a truly global scale, tropical forests suck carbon dioxide out of the air, which is important to us all because carbon dioxide is a key greenhouse gas responsible for global warming. Tropical forests are thus known as "carbon sinks," and their vegetation alone stores an amount of carbon equivalent to more than half of what is in the atmosphere.[5]

Yet the boreal forests hold even more carbon than tropical forests, and the tropics have a longer growing season for agriculture. So can we swap ecosystem functions and cut down the tropical forests? In other words, do we need ecosystem diversity? Let's think about the consequences of cutting down the tropical trees. First, the earth would lose a large carbon sink that helps ameliorate the greenhouse gases emitted by industrialized nations. If we burned the wood to free the nutrients for agriculturally rich soil, we would be releasing most of the stored carbon into the atmosphere as carbon dioxide, thus exacerbating the problem. The increased atmospheric carbon dioxide may help our crops grow so that we can better feed our growing population, but those of you who live on the coasts will have to inch inland as the seawaters rise.

Ecologists look at not only how climate change may affect biodiversity, but also how plant species collectively affect weather patterns and climate. Forests absorb and utilize heat and water. Bare soil responds to the elements quite differently. Denuded patches of earth heat up and desiccate, resulting in a localized area of dry heat that can disrupt regional weather patterns. In chapter 7, we saw how this phenomenon contributes to desertification in the Sahel. Back here in Ohio one can witness a comparable phenomenon created by denuded urban landscape. Everybody loves to talk about the weather, so I often watch the Doppler radar screens on TV as rain clouds pass over Columbus. As the dry heat rises from the city, it often dissipates clouds, and on the radar one can see them break up as they pass (fig. 8.1). It looks as though Moses were standing at the west side of town, raising his staff to part the waters. The same effect can occur in the deforested tropics, with a net loss of rainfall for agriculture in those regions.

The plants of the forest transpire moisture into the air, which then accumulates as rain clouds. Although invisible to us as it is happening, this lifting of water is not inconsiderable: a single tree in the rain forest returns about 2.5 million gallons of groundwater to the atmosphere over a hundred-year

FIGURE 8.1 Landscapes that are cleared of vegetation for cities or agriculture absorb and radiate heat in predictable ways. This can disrupt local weather patterns and may affect climatic patterns elsewhere. (*Illustration by Michael Masters*)

lifetime.[6] Without trees and plants to cycle water, places such as the Amazon Basin would lose a significant amount of rainfall, as would other parts of Brazil where crops are grown. There would be fewer storms to drench young naturalists following in Darwin's footsteps, and less biodiversity for them to find as well.

Weather around the world is governed by global currents of both air and seawater, which are in turn influenced by natural ecosystems. Altering those natural systems disrupts the currents in unpredictable ways, much in the way the El Niño warming of the Pacific Ocean ultimately affects the rain in South Africa. The removal of a tropical rain forest may not keep it from raining in London with monotonous regularity, but would in some way alter England's green and pleasant land. Best we keep the rain forest intact.

Home on the Range

Ever since one of our early ancestors ventured out of the forest and strode across the African savanna, grasslands have been an important part of the human ecological niche. Although anthropologists have largely abandoned the idea that our upright stance was specifically adapted to the savanna environment, the fact that our ancestors survived in such environments fueled our evolution and opened up new corridors for dispersal. By ten thousand years ago and thereafter, grasslands provided grazing land for the first domesticated animals, and seeded our gardens with the first crops. Corn and wheat, after all, are grasses.

Grasslands in the temperate regions have been very inviting for agricultural conversion, as they were to the first agriculturists. So if we are to leave the forests alone for their ecological services, and grasslands have natural affinities for agriculture, should we worry about conservation of the natural grasslands that remain? Or do they hold some intrinsic value that cannot be measured in dollars and cents?

One economic benefit of natural grasslands is that they are breeding grounds for birds and bugs. Oh, great! Birds to eat people's grains, and insects to infest them. But one must remember that our crops are just an adaptation of natural ecosystems, and still utilize processes evolved from before the time of agriculture. Many of the wild bird and insect species help our crops, either as natural pest controls or as pollinators. Taking them away would leave the farms unmanageable, and the cost would be enormous.

Perhaps the greatest value of natural grasslands is that they hold a "genetic library" for the grain crops that feed us (and that also feed our domesticated animals).[7] Whereas wheat and corn have successfully adapted to human populations, they did so at the price of genetic variability. Our monocultures of crops are susceptible to blight from a range of pesky diseases and pests that can evolve rapidly; thus the crops could do better with a more varied genetic endowment. Whereas crops that are genetically modified by insertions of genes from other life-forms can increase yields, many people have concerns about their potential environmental impact and unknown effects on human health; they don't want to suffer the potential consequences. But the genetic ancestors of these crops, some of which still live in wild grasslands, hold more "natural" genes that can infuse our inbred crops with greater variability.[8]

Like forests, natural grasslands also provide a carbon sink. They differ from forests in that most of the carbon is stored in the soil.[9] Likewise, natural grasslands are more effective than farmed fields at absorbing rather than reflecting solar energy. The extensive root systems also reach down into the soil and tap water that is then transpired into the atmosphere. Thus the world's climatic regime is held in check, at least in part, by the natural services of grassland ecosystems. Moreover, because of the complex interactions of roots, microorganisms, and other life-forms in the soil, tall-grass prairies acclimatize readily to global warming without accelerating the respiration of carbon dioxide, whereas other ecosystems tend to exacerbate atmospheric carbon dioxide loads with increased respiration under warmer temperature conditions.[10]

Through tilling the ground and planting monocultures, or through overgrazing and trampling by too many cattle, farmlands experience a great deal of erosion. Not so in natural grasslands, which have perennial root systems that nurture the soil and prevent its erosion. Indeed, the wild grazing animals such as bison help to maintain the system by stimulating the growth of new shoots, and by leaving deposits of natural fertilizer. Recall that the loss of North American grazing megafauna had an effect on many species in the former ecosystems because of the loss of their services in maintaining a particular mix of vegetation. Likewise, prairie dogs, gophers, and other burrowing rodents recycle nutrients and maintain the best fodder for the grazers.[11] A prairie is a product of the diverse activities of many species.

But a growing population of hungry people will ask: what good is that natural soil to us if we can't use it? Aside from being a carbon sink, not much, although a prairie dog might beg to differ if it could. On the other hand, grasslands converted to agricultural use promote soil erosion and other consequences far from the farm. "Off-site erosion costs include expenditures such as the increased costs of obtaining a suitable water supply, maintaining navigable channels and harbors, increased drainage problems, increases in flood damage, increased costs of maintaining roads, and decreased potential for water power."[12] If you want that in dollars and cents, it comes to the tune of $17 billion a year in the United States alone.[13] The prairies left intact cost us nothing.

Find the River

"The rivers are our brothers. They quench our thirst. They carry our canoes and feed our children. So you must give the rivers the kindness that you would give any brother." So go the words attributed to Chief Seattle of the American Northwest Coast's Suquamish tribe.[14] Rivers and lakes have always played a key role in human survival, despite making up only a small percentage of the water on earth. Not only do they furnish us water for drinking and irrigation, they also serve in recreation and transportation. Of what practical benefit is the *biodiversity* that fills those waters? After all, we can drink from a tap, swim in a chlorinated pool, and paddle our canoes across lifeless waters.

Just don't try to go fishing. Freshwater ecosystems first of all supply us with food, as do ocean ecosystems. That function alone should be sufficient to warrant conservation of such ecosystems. The web of life from phytoplankton up to the top predators depends on the biodiversity throughout, and an overindulgence of human interference puts our food source at risk.

But the value of life in the water of rivers and lakes goes beyond our food supplies. For one, the waters feed other natural ecosystems. Just go to a quiet spot where the forest edge meets the river, as I often do, and watch the interactions. Wild animals come down for a drink, and some dine on the mussels, crustaceans, and fish. Birds will swoop down for a taste of the insects that hover above the water. Even the trees extend their roots into the rivers for water and nutrients, meanwhile helping prevent erosion of the banks. If we value the forests for their services, and grasslands too, then we need the aquatic ecosystems that supply them with food and fresh water.

Plant and animal species in the rivers, lakes, and oceans also help to clean up after us, as long as we are sparing about how much of our waste we send their way.[15] Aquatic life gets right to work on our waste, transforming it into nutrients for the food chain, detoxifying some of the pollutants we spill, and sequestering away others in sediments below the water.

What about swamps and marshes? They stink. They are also full of life that is very effective at cleaning the water. Furthermore, when floodwaters rage down a river, swamps can absorb the overflow and save the cities downstream from destruction. When storms pound the ocean shore, it is the marshes and estuaries that again save human civilization, for their plant roots and networks of burrows hold the land and absorb the shock. In New Orleans, Louisiana, there is little protection left from hurricanes because artificial structures have altered the natural buffering effects of the Mississippi Delta. Even without a major storm, the Louisiana coast loses one acre of land to the encroaching sea every twenty-four minutes.[16] The loss of the marshes may have huge economic costs, and lead to substantial loss of human life should New Orleans get a direct hit from a hurricane.

It is not just the water we need, but the life in the water. Chief Seattle apparently knew that as well: "Man did not weave the web of life, he is merely a strand in it. Whatever he does to the web he does to himself." [17] The good chief apparently had a prescient knowledge of ecology, decades before the science was formally born, and a century before the concept of the "web" came to dominate ecological science. It just goes to show what we can all learn if we take the time to observe the world around us.

South African Fynbos Revisited

So far we have looked at the practical contributions of major ecosystems. But why worry about minor ones such as the Cape Floristic region, home of the richly diverse fynbos, even if it does have a large number of endemic species? Sure, it is pretty to look at and thus has tourist potential, but as one of the hottest of the "hot spots" of biodiversity targeted for conservation, we should see some justification for the effort. It's there if we just pause and think. [18]

Well, prettiness and tourist potential are legitimate concerns—they put a practical price tag on Elton's aesthetic reason for conservation. In a country such as South Africa, which was struggling economically at the turn of the century, "eco-tourism" is an important part of recovery. Moreover, the beautiful fynbos wildflowers are harvested, cut, and dried for floral displays, and have fueled a thriving enterprise.

The fynbos plants do more than that. One must understand that they are superbly adapted to the infertile ground, wet winters, and dry summers. The invasive plants have adapted to such conditions as well, but do not have as sparing a nature when it comes to water consumption. Water running off the mountains is fed into the sparing fynbos vegetation, which cleanses it for the human population (and other living things) in the region. An invasive pine forest might provide lumber and fuel, but it sucks up the water that gives life.

Every ecosystem, like every person, has its role to play in sustaining a livable environment for us all. But having taught human anatomy for many years, I'd like to dissect the ecosystem "bodies" further to help understand the roles of biodiversity.

Intertwined Strands of the Web

Ecosystems are complex functioning units, just like human bodies. [19] Every body part, like every species, has a role to play in keeping the system alive and healthy. Our bodies have many redundant parts. You can lop off one finger and still have nine left, or do a Van Gogh with your ear and still hear reasonably well.[20] Many key organs are duplicated, such as the eyes or the kidneys, providing a backup in case of failure. But failure of one eye or kidney places additional stress on its alternate. Likewise, the failure of one body part may put stresses on a seemingly distant organ. For example, the breakdown of valves in the veins of the leg (that breakdown gives us varicose veins) may put additional stress on the heart, which has to pump harder to ensure that the blood flows back up the leg. Or an infection of one part of the body may be carried through the bloodstream to another.

The web of life suffers from comparable maladies, and is best maintained by duplicated and interacting systems — that is what makes species biodiversity so valuable. There are evolved redundancies among ecosystem species, or so it would appear. Multiple species that serve the same function spread the risk should one species fail to perform its task. This makes the ecosystem more sustainable, or at least more resilient as the world changes from year to year. But over time, those apparent redundancies become less real.[21] No two species are exactly alike, and even slightly different species can respond to change in different ways.

How does nature tolerate even partial redundancies of species? It would seem that in Darwin's world, one slightly better-adapted species should outcompete the other, leaving the ecosystem to function just as well with less biodiversity. But there is also the side of grandeur in evolutionary ecology, and that is well illustrated by two monkeys in southeastern Peru.

Saddle-back tamarins and emperor tamarins are fairly similar species, and for the most part live on the same fruits (though the former has a greater penchant for insects).[22] Over the course of a season, the fruits of small trees and vines ripen gradually, and are eaten by both species. But rather than rush off and secure an area for themselves, the two species share territories and even call to each other to keep in contact. This mutual relationship hardly seems productive from a superficial perspective, but it works well due to a number of advantages it affords.

One bonus has to do with predator protection. Both tamarin species are constantly on the lookout, and they quickly respond to each other's warning calls. This alone does not explain their cooperation in other realms, but is an important component of how the system evolved. When two species do the job of one, it decreases the disruption to any one of their populations by spreading the risk of predation across the two species. Each is an equal target.

The sharing of food resources is even more intriguing. Each tree or vine is harvested no more than once every few days. The tamarins regulate the time between visits, thus allowing more food to ripen. Why not compete for food? Without cooperation, they would not know when a tree had last been harvested, and would make many disappointing and inefficient trips to trees from which the ripe fruits had already been snatched. By moving together, they ensure that each trip will be "fruitful." It's rather like following the crowds to find the best restaurants in order to avoid the substandard diners.

Between efficient resource sharing and mutual protection, the two tamarin species get along better than either species could do on its own. In other words, there are payoffs to the apparent redundancy, and evolutionary reasons for the species biodiversity. Furthermore, as time goes on and the environment changes, subtle differences between "redundant" species may allow divergent adaptations to the new regime.

Leave It to a Beaver

Of course ecosystems involve many unrelated species that seem to go about their businesses without much consequence to or reliance on the others. But as in the human body, all parts are interconnected — sometimes in unexpected ways. Ecologists have documented the intricate "bodily functions" of many ecosystems, and as an example I'd like to take you back to the Boundary Waters Wilderness. The late Miron Heinselman eloquently pieced together a fine example of the interactions within an ecosystem, and how the illness of one species leads to stresses on others. Here I will try to do justice to his work, although in shortened form. Heinselman called it the "Moose-Deer-Caribou-Beaver-Wolf system," but he made it clear that other species are involved, including plants and invertebrates, but most notably humans and brain worms.[23]

Ever eat a pine tree? Many parts of it are edible, especially if you are a moose. A moose can live on leaves, twigs, pine needles, and vegetation from lakes, largely with the help of microbes in the first of its multiple stomachs. The moose is an enormous animal that one can hardly miss, but I did miss seeing them during my boyhood adventure in the Boundary Waters Wilderness, despite canoeing across the lakes and camping in the tree stands they prefer. It seems that at the time their populations were low and just beginning to recover from the devastation of a parasitic brain worm.

The tiny parasite that so effectively brought down the massive moose had an interesting point of entry: the snails and slugs that hosted the parasite would be inadvertently consumed during the moose's unrefined munching of plant material. It turns out that the snails and slugs had picked up the brain worm larvae from their own close encounters with the fecal pellets of white-tailed deer. Unlike the moose, the deer are unaffected by the parasite.

Whereas white-tailed deer are not really alien invaders of the Boundary Waters Wilderness, they are at the northern end of their habitat range there and would not normally be found in large numbers. Deer prefer disturbed areas to mature forests, so their habitat became increasingly abundant with more settlements and logging activities as the human population pushed through the area, starting around 1890. The deer population blossomed, and with it the population of the brain worm they carried.

Eventually the disturbed area went through successional stages to become mature forest again, which was as well suited to deer preferences. Deer, however, were to the liking of both the wolves and the humans who hunted them. Moreover, heavy snows during some particularly harsh winters made them more vulnerable to wolf predation (whereas moose and caribou were better adapted to maneuvering through the thick blanket of snow). So the deer population collapsed rapidly back to more normal levels, but significant damage had already been done to the moose and caribou.

Woodland caribou are impressive animals that are well adapted to the cold and snow. They fill a niche between that of the deer and the moose, and thus are an appropriate meal for wolves, lynx, and bears. I never got to see any caribou either, and you probably cannot see one in the area today. They too succumbed to the brain worm brought by the deer, a situation exacerbated by excessive hunting by humans.

Enter the beaver. Beavers play a key role in the local ecosystem through their peculiar habits. They build lodges in the water, where they are safe from predators year-round. The occasional wolf gets them, to be sure, but wolves are better fed with the copious meat of larger mammals. The beaver lodges also serve as a place to store food for the long winters. In order to have enough space under the ice, and for increased security, beavers also build dams to raise the water level. This is great for those of us who love canoeing, for at lower levels some lakes are impassable. More important, the beavers' work also expands the lake habitat for waterfowl and moose.

Enter the human, again. Beaver populations were decimated in the Boundary Waters Wilderness and elsewhere by the fur trade. To further compound their woes, beavers suffered a fate similar to that of the Idaho ground squirrel (chapter 6) due to human-enforced fire suppression. Without fires, the pine trees start to replace the aspens and birch trees needed by the beavers.

Without beavers around to perform maintenance duties, their dams eventually broke down and released the water. This was too bad for the moose population, which was on the rebound after the collapse of the deer population. To be sure, in some parts of the region, moose had become so successful that they needed to be culled by humans (the ecological transition revisited). But in other areas their habitats had shrunk due to the loss of the beavers and their dams. On the other hand, the moose were now targeted more and more by wolves. Wolf packs vigorously defend their territories, hence their awesome howling episodes. But with the loss of the deer and caribou, wolf packs became stressed and had to shift territories, as well as dietary preferences — back to the precarious moose population.

There is no telling how this story will continue to unfold. Like the human body, a dysfunction in one part leads to illness of the entire system, and the ecological doctors have a difficult time finding a cure. That is not to say that shifting populations are not normal in nature; evolution and extinction would not work otherwise. There is no such thing as a "balance of nature," unless one thinks of it as a dynamic balance, just as nobody is free from disease or eventual death. But the story does show that many species make up the web, and that humans are merely one strand in it, thus sitting precariously in a position that Chief Seattle apparently knew well.

Keystone Cops

Whereas many parts of the human body have backup systems or duplicates, some do not. Sometimes these serve key functions, such as the heart or the brain. A space shuttle has multiple computers to serve as backups, but humans have just one computer each, making it a very important part to maintain. If the brain or heart gets sick, the whole body suffers. If either ceases its function, the body dies.

It has been argued that ecosystems have particular species that serve key functions, and they are known as "keystone species." Whereas we have just seen that many microbes, plants, and animals make up a functioning ecosystem, in some cases one species plays a disproportionately important role, as the metaphorical heart of the ecosystem. If that one species dies, with no backup species to take its place, then the whole ecosystem decays or dies.

Keystone species come in many shapes and sizes, and can play a variety of critical roles in ecosystem functioning.[24] Oddly enough, keystone species are often the predators in a system. One might think that other species would be better off without being eaten, but the "balance" of the system depends on their numbers being kept in check—sort of a Malthusian principle of the wild. The sea star, for which the "keystone species" term was developed, is an example of such a predator.[25] Without the sea star, the numbers of other species in the marine intertidal ecosystem actually drop dramatically. The sea star feeds on many of the invertebrates such as mussels, barnacles, and snails. But in its absence, one mussel is able to expand its population to take over the limited available space in the intertidal zone, to the exclusion of other species. It should not escape your notice that mussels sound rather like us humans, who currently have no major predators to keep our numbers down.

Another classic example of a keystone species is the sea otter, relative of the cute otters I saw in the Boundary Waters Wilderness. Sea otters consume great quantities of sea urchins, and keep their numbers in check. When the fur industry and incidental effects of fishing decimated the sea otter population, sea urchins had a population explosion and overgrazed the algae and kelp.[26] The destruction of the kelp forest destroyed the habitat for many other species, until sea otters were reintroduced and the recovery process began.

Keystone species need not be top predators, however, as other types of species are vital to maintaining ecosystem biodiversity. For example, there are "ecosystem engineers" such as the beavers we just looked at, who help expand the habitats of moose and waterfowl.[27] Beaver dams also increase biodiversity from lake to lake in two ways. By limiting the exchange of fishes between lakes, the process of homogenization is avoided. Meanwhile, the enlarged lakes expand the fish habitat for larger populations to conduct evolutionary experiments in adaptation.

The problem is that keystone species are difficult to identify. They could be anything from the tiniest bacterium to the moose. Which one in the Moose-Deer-Caribou-Beaver-Wolf system would be a keystone species, or would the aspens and brain worms also qualify? Sometimes the loss of one subtle keystone species may have an effect only generations later, as the consequences of its demise work their way through the web. That is why maintaining high biodiversity is important to maintaining the ecosystems that serve us so well.

For those keystone species that can be identified, conservation programs can gain focus and relatively quick success, such as with the reintroduction of the sea otter. So it is worth pursuing a clear working definition of keystone species. The current scientific consensus definition is this: "A keystone species is a species whose impact on its community or ecosystem is large, and disproportionately large relative to its abundance."[28] Thus, say, the abundance of baboons or impalas in South African environments does not make them keystone species, whereas a few hummingbirds may be vital to the pollination of trees that feed many a mammal.

Here I would like to break with convention and nominate human beings as keystone species. Even when humans were not as abundant as we are today, they've had a disproportionate effect on any community or ecosystem in which they've lived; that much is clear from the fossil record. Like many keystone species, we are a top predator (although we don't need to be). With the ecological transition, we also clearly qualify as ecosystem engineers,[29] in modes ranging from the transformation of lands for agriculture to the culling of moose in Canada. And now biodiversity depends on us in many ways.

What I wonder, however, is what happens when a keystone species with a disproportionate effect becomes *more* abundant. Nature often has checks and balances that prevent, for example, sea otters from becoming too numerous. Humans keep on sidestepping those population checks, making us a very unusual keystone species. In our bodily analogy, human population growth can unfortunately be likened to the spread of cancer cells. As the cancer cells initially grow and multiply, they more strongly affect the functioning of an organ. As they become even more abundant, they eventually shut down the organ. If they spread to other parts of the body, they can shut down entire systems that keep the body alive. Likewise, ecosystems that keep the planet alive suffer under the cancerous growth of the human population.

I think that this is the most depressing analogy I've ever devised, though it has an unfortunate ring of truth about it. On the other hand, unlike cancer cells, we have the power to be cognizant of what we are doing to our host. We can perceive our position and act accordingly before it is too late. That is the grandeur of being human. If we are to follow through with the ecological transition and manage ecosystems by reintroducing some species or culling others, perhaps we should consider managing our own species before a human "cancer" consumes the world body.

Biodiversity and Productivity

A recurrent theme throughout this book has been that high species biodiversity is more productive than low biodiversity. That is why we can look at infrared satellite images and identify the most diverse areas of the planet—their productivity is evident in the energy patterns revealed. We know our monocultures are less productive and more prone to soil erosion than more diverse ecosystems, yet we persist with the practice because it focuses the productivity on foodstuffs we need or desire. But perhaps the greatest value of biodiversity is in sustainable productivity.

The productive value of species biodiversity is not new ecological knowledge. Decades before there was a science of ecology, Charles Darwin postulated that more diverse plant communities had greater productivity.[30] Decades before the term "biodiversity" was coined in 1986,[31] Charles Elton—in the same work where he discussed the three reasons for conservation—proposed that more diverse ecosystems were more stable (and thus sustainable). Modern ecologists have increasingly sophisticated tools to test these ideas, as well as rigorous debates about what the tests mean.

Clearly on the large scale, across continents and major ecosystems that we view with our satellites, biodiversity and productivity increase together.[32] It takes many different species to best utilize the variety of resources across various substrates, the uneven water distribution due to varied rainfall patterns, and so on. Thus the relationship between biodiversity and productivity is partly related to the amount of area covered—size matters.[33] One set of species cannot effectively do it all.

On smaller scales the nuances and complexities of ecological systems start to show. One cannot just cram a whole bunch of species into a given area and expect increased productivity. Too many species, or an incoherent set of species, may actually decrease productivity.[34] This is one of the reasons that habitat fragmentation threatens biodiversity, for small units may not have the diversity of terrain for all the species we try to preserve there in their "natural" environment. In other words, each area we isolate from the others is going to have its own optimum number of species to ensure productivity, and it is difficult for science to predict what that number should be, or which species make the system work.

A fascinating and controversial set of field experiments conducted by a team led by David Tilman of the University of Minnesota illustrates both the principles and the complexities.[35] They tested a number of grassland plots in which they controlled the number of plant species for each plot, 147 plots in all. They found that plots with more species were more productive (i.e., produced a greater bulk of plant material, or "biomass"). Moreover, the plants in plots with greater numbers of species better utilized the limiting nutrients in the soil (in this case the soil's mineral nitrogen). Thus they concluded that their experiments supported both Darwin's and Elton's notion of the positive relationship between biodiversity and productivity/sustainability.

The critics of these experiments have gone so far as to call them "irrelevant." [36] They suggest that the results depend not just on how many species, but on *which* species are present. [37] The more species you have in a plot, say the critics, the more likely it is that one of them is going to be productive. Whereas this may be true, isn't that the way it works in nature? Not every species in a natural ecosystem is equally productive, and indeed some ecosystems can lose species with little ill effect — that's the way it works in Darwin's world. High biodiversity does ensure that the best species will be there, and natural selection does the rest. Thus the critique is specious at worst, and at best shows the complexities of nature pitted against the limitations of a young science. Moreover, further experiments have yielded consistent results, which along with mathematical tests provide further evidence for the productive benefits of biodiversity due to complementary species interactions. [38]

Fortunately, scientists on all sides of the debate are coming together to resolve their differences. Although there is more work to be done, their current conclusion is this: "There is consensus that at least some minimum number of species is essential for ecosystem functioning under constant conditions and that a larger number of species is probably essential for maintaining the stability of ecosystem processes in changing environments. Determining which species have a significant impact on which processes in which ecosystems, however, remains an open empirical question." [39] So no matter how you look at it, species biodiversity is still at the heart of ecosystem sustainability.

None of these studies, or counterarguments, is inconsequential, as environments *are* changing. One effect of the growth and industrialization of our human population has been the increase of carbon dioxide in the atmosphere. One may argue about what if anything that means in terms of theory and climate models — the "greenhouse effect" and global warming — but since 1870 the amount of the gas has increased in our atmosphere from 270 to 370 parts per million. [40] Those are data points, not theories. The good news is that the carbon dioxide increases plant productivity — although probably not by much. Fine, if you don't live on a coast. But once again, species biodiversity comes into play if we want to tap into the productivity bonanza, whatever its size, or capture more carbon from the atmosphere to buffer global warming.

One of David Tilman's colleagues at the University of Minnesota, Peter

Reich, led a team that studied the effects of carbon dioxide on grassland plots harboring various levels of biodiversity.[41] They artificially raised the levels of the atmospheric gas around the plots, which had either one, four, nine, or sixteen species of plants. You can guess which plot was the most productive.[42]

At this point we have found practical value in ecosystem biodiversity for everything from carbon storage to water cleansing. Within ecosystems, and within subsets of ecosystems, species biodiversity is necessary for the health and productivity of the web of life. But if we want to keep those species alive in our ever changing world, or allow new ones to evolve and adapt to the emerging conditions, then genetic biodiversity becomes important as well.

A Dip in the Gene Pool

As wild plants, animals, and microbes retreat into the remaining fragments of their habitats, there is clearly a loss in their population numbers. With that loss of numbers comes a termination of many family lines within a species, and the extinction of the peculiar genes they carry. In other words, there is a significant loss of genetic biodiversity.

We tend to think of extinctions in terms of species, but the loss of wild genetic variability is at least as important for sustaining our planet and ourselves. It is difficult, as we have seen, to count species extinctions. Quantification of genetic extinctions is currently beyond our reach, and such numbers may be inestimable. But it is not the numbers that are most important; it is understanding the principles of genetic losses. By knowing the principles, we can understand the consequences.

Usually the practical benefits of genetic biodiversity are broached in terms of their economic benefits for agriculture (discussed earlier) and medicine. For example, more than half of all prescription drugs are modeled on natural compounds, and about a quarter are taken directly from plants (or are modified versions of plant substances). Next time you pop an aspirin into your mouth, think of it as an example.[43] The wealth of unexplored plant species holds significant potential for more discoveries like aspirin. So far, about fifty thousand plant species have been screened for medicinal compounds, and they have yielded about fifty beneficial drugs. So for each thousand plant species that go extinct, we may lose one or more pharmaceuticals.

It is not the plant species per se that provide us with our medicines, but the chemical compounds produced by the genes in individual plants. Plants within a wild species may vary considerably in their genetic makeup. Not all members of a species may produce the particular compounds that could help us cure cancer (though we already have a cancer medication from periwinkle flowers) or battle current and future epidemics. So genetic losses can in some cases be as consequential as the loss of a species.

Although this book has focused on the practical benefits of biodiversity, human cultural diversity often comes into play as well. I'd like to take a brief foray into the value of cultural diversity in order to make a point. Increasingly the world is dependent on Western-style medicine, but many non-Western peoples take care of their ills with native plants. Through years of experience and experimentation with their natural environment, carried on from generation to generation, they have built up a great store of knowledge as to which plants can cure diseases or alleviate symptoms. It is not magic or superstition, but exploitation of biological compounds. As we assimilate those people into our growing Westernized population, those cultures become more homogenized, and we are at risk of losing the knowledge and wisdom they carry about nature.

Native Americans provide such an example, and lead us back to the value of biodiversity. Not only were they good at breeding corn, but they knew medicinal plants. One of them is known as "bear root," which has chemical compounds that treat a spectrum of ills from headaches to fungal infections.[44] The name of the plant, however, gives away the source of the Native Americans' knowledge: according to Navajo legend, the bear gave them the plant as a gift. Indeed, it is now known that bears will chew it up and rub it over their bodies, giving some credence to the myth.[45]

Thus other animal species also have knowledge, or at least evolved experience, relevant to finding and testing medicinal compounds. We can learn from monkey see, monkey do. In the tropical forests of Venezuela, capuchin monkeys rub themselves with a particular type of millipede. Now millipedes carry a lot of chemical compounds, but researchers have found that the apparent benefit comes in a particular chemical that repels insects, an important consideration for the capuchins—and humans—during the rainy season.[46]

Many animals have such lessons for us. We can observe the ways they use their natural resources, how plant and bug juices interact with their own genetic makeup, and how we might make use of the chemical compounds ourselves (if we are genetically similar enough to the animals to benefit in comparable ways). But for that we need both the plant and animal species, and the diverse genes they carry.

Safety in Numbers

We have already seen how diversity among plants makes our crops more resistant to disease than inbred crop monocultures. For example, experiments in China compared the yields of genetically diversified rice crops with those of genetically uniform monocultures. The mixed rice crops had 89 percent greater yield and were 94 percent more resistant to the blast disease to which some varieties succumb.[47] This is important to us because rice is the staple crop for about half of the world's population.

The practical principles of genetic biodiversity apply in nature as well as agriculture. Smaller, inbred stands of natural plants have less variability to withstand an onslaught of disease or pests because their natural resistance is severely diminished. The same is true of animals as well. If there were still a large, thriving population of gorillas, some would naturally have survived an epidemic of measles; but with just a few members of the species, conservationists have had to resort to inoculating the gorillas so they could survive. Obviously, we can't provide that level of protection for twelve million natural species. Only genetic variability can give them the tools to maintain their populations.

Loss of genetic biodiversity has further long-term consequences. Over the course of earth's long history, the slow process of evolution has regenerated species biodiversity after extinctions large and small. But since evolutionary origins are slower than extinctions, it takes about ten million years for biological recovery.[48] True, some species evolve more quickly than others. Plants of the Cape Floristic region apparently evolved "rapidly" over the course of no more than seven or eight million years.[49] That is still a long time coming. For the species replaced, it will be an even longer time gone.

So already the human impact has had tremendous consequences, the resolution of which none of us who are alive today will ever see. But if we persist in fragmenting habitats and diminishing the genetic variability of species, the survivors will continue to become less resilient and less evolvable. There will simply be too few gene variants for natural selection to act upon. Given the exacerbation of species biodiversity loss with the depletion of genetic biodiversity needed for the evolutionary generation of new species, the inevitable result will be further delays in recovery. As Harvard scholar E. O. Wilson put it, "humanity has already closed most of the theaters of natural evolution."[50] As long as our population continues to grow, or even if we maintain our current population levels, those theaters will not reopen.

Genetic diversity is as important to the survival and sustainability of the living world as it is to our bodies. Each human being is made up of over two hundred types of cells. Skin cells, muscle cells, brain cells, blood cells, and the rest are ultimately constructed and diversified by chemical programs held in the DNA of an estimated thirty thousand genes in our genome.[51] We use all those cells, and they depend on a diversity of genes. So do species, the organs of the ecosystem body. If we continue the analogy configured throughout this chapter, bodies (or ecosystems) depend on the diversity of organs, most of which must function together for life to persist, and all of which lead to good health. Each person, like each ecosystem, has his or her own occupation that in concert with others makes the world go 'round with life. It is up to us to make sure that the component parts stay healthy so the whole does not die.

CHAPTER

Epilogue: The Keystone Species with a Choice

EACH weekday morning, before I get down to the day's writing and research, there is a fairly normal routine to follow. By the time I get cleaned up, dressed, have a chat with my wife, eat breakfast, dress my two boys and get them off to school, about an hour and a half has elapsed. For some people it takes that long just to commute to work; others have routines comparable to mine. The time always seems to go by quite quickly, unless one is stuck in traffic, but meanwhile the fast-paced world of ours churns on.

Here is something to think about during your morning routine. In just that hour and a half, on average, 22,530 babies are born. Hopefully every one of them is loved and cherished. Meanwhile, 9,418 of our loved ones pass away, each having played a part in making the world what it is today. That means a net gain of 13,112 people — quite a few for such a short time span.

That same amount of time also sees the end of at least one entire species, by the most conservative estimates. The last individual gets eaten, dies a quiet death, or gets cut down in the name of human progress and expansion. Chances are that the number is more like seventeen species gone extinct in the hour and a half. Meanwhile, it is unlikely that a new species has evolved. Perhaps a new species of bacterium arose, but on average each morning it appears that there would be a net loss of species biodiversity. Also lost is what the species did to maintain the world's ecosystems. There is no telling how many genes of other diminishing species were lost during the same time, as fewer individuals survive the morning.

The gains in the number of people and the loss of other species are related. To be sure, the number of lost species is mediated by the choices we make every day as a keystone species. We can help their survival by simple considerations about how much water we use to brush our teeth or by using public transportation rather than our cars. Indeed, some people have chosen to be proactive and work each day toward conservation of the remaining habitats of those animals. How we act thus leads to some of the proximate causes that determine the life or death of a species in the wild. But the underlying cause of their decline ultimately comes back to how many people there are, and how fast our population is growing. The more of us there are, the fewer of them there can be.

There are those who see human population growth as a good thing. More people mean more ideas on how to avoid the dire predictions of Thomas Malthus and sustain a growing human population. By using better and better technology, says cornucopian eco-optimist Ronald Bailey, "humanity has avoided the Malthusian trap, while, at the same time, making the world safer, more comfortable, and more pleasant for both larger numbers of people as well as for a larger proportion of the world's people."[1] It is true that the past century has seen many advances in agricultural techniques and energy extraction that have made the world more pleasant for *some* people. Then again, Bailey's "man must conquer nature" attitude has led to other ideas such as Mao's policy of killing the sparrows or the Americans' introduction of multiflora rose — all well-intentioned, all disastrous. Is there really safety in numbers?

Remember TANSTAAFL: there ain't no such thing as a free lunch. In the words of my doctoral mentor, Stephen Molnar of Washington University: "We pollute, extract, destroy, and displace people from adaptations that had sustained them from hundreds or thousands of years. The quest for a means to expand the earth's human-carrying capacity has been costly, with many casualties along the way."[2] Biodiversity is one of the casualties.

Many species are already extinct. There is nothing we can do about that now. And given the extinction debt — species already on the wane who have lost the genetic wherewithal and habitat expanse for recovery — it is inevitable that many more will go extinct over the course of the next century. Meanwhile, the relentless growth of our human population appears set to continue unabated for some time. Even if we immediately went to replacement rate (two offspring per couple), the world's population would continue to grow due to the population age structure and declining mortality. Both extinction and population growth are natural phenomena. What is unnatural is the magnitude of both trends in today's world. And what is unusual is that one species has a choice to alter the course of things to come.

If we look down the road toward the long-term sustainability of both people and species, there are two main measures that we can take — and must take now. One is active conservation of the natural world; the other is proactive population control. I hesitated when I wrote the word "control," for we must respect the rights of people to reproduce. But both conservation and population growth abatement, if you prefer, involve a global conscientiousness that can be achieved only through education. That is the job, indeed the duty, of everybody who reads this or similar books: pass on the word.

Simple Conservation

Turn off the lights. Turn down the thermostat in the winter. Turn off the tap when you brush your teeth. Recycle waste. Carpool to work. Eat less meat.

Here in the United States, the most rampantly consuming nation on earth, such proselytizing gets irritating. But these simple measures would mean a lot. Many Americans think that such practices would be an imposition on their lifestyle. They really would not. In many other places, such behavior is a subconscious way of life, even for those living in opulent style. It just takes a bit of practice at first, and then it becomes second nature.

It is amazing what a bit of simple conservation can do. I often watch people in the parking lots at shopping malls drive round and round looking for the closest available spot to the door. I've done some rough math and found that if Americans would just park sooner and walk a few extra yards, in one year they could save enough fuel to supply *all* of South Africa's petroleum needs for nearly a full workweek, or Zimbabwe's needs for two months, or the country of Georgia's fuel consumption for more than a fifth of a year.[3] It would mean as well a little less frustration in the parking lot. It is not hard to do, and the economic benefits would help everybody live a more opulent lifestyle.

What would these things do to help other species survive? Watching water consumption and waste disposal would give them more and cleaner water to drink or swim in. Cutting electrical and gas consumption would provide cleaner air, and stem the disruptive effects of rapid global warming. With less energy consumption, we'd be less prone to intrude on wilderness areas in search of fossil fuels.

But "environmentalist wackos" cannot cajole everybody into taking such measures. People will do as they please. It has evolved in our psyche to look after one's self and family — that is part of what has made us so successful in Darwin's world. We have also evolved to see the here and now, rather than the grander picture. Moreover, our cultures have thrived on individual freedoms. So grander measures must be taken to ensure a sustainable world in which we can continue to reap the benefits of biodiversity's natural services.

Complex Conservation

Conservation of wilderness lands is a noble goal, and much has been achieved through the establishment of reserves and parks, as well as attempts to restore natural ecosystems. It all sounds so simple and easy: set the land aside and let nature take its course. We reap the benefits of ecosystem services and preserve higher levels of biodiversity in the process. Without conservation, we are basically outcompeting not only other species but our own human descendants for the values that biodiversity bestows.[4]

But the situation is not simple and easy; indeed, it is quite complex. Ecology is a young science struggling to come to grips with some of the most complicated research questions on earth—how ecosystems work. This involves bringing together data on plant and animal taxonomy, evolution, anatomy, physiology, behavior, genetics, biochemistry, and more with understandings of soil substrates, weather patterns, atmospheric gases, and so on. Great strides forward have been made, and every week new research is published that gets us closer to greater comprehension of the living world. But there is a long way to go, and that puts conservation programs in a bind.

Much of the land our civilization has reserved for wildlife is just a pittance compared to what formerly existed, and is grossly fragmented. In order to maintain wildlife habitats in a functioning capacity, we have to intervene. This is the downside of the ecological transition, because now we must take care of not only our agricultural and urban communities, but natural communities as well.

So conservationists are forced into intervening, and answering vexing questions. Do we cull moose in one part of the Boundary Waters Wilderness or reintroduce beavers in another part to reestablish the moose habitat? Do we cull elephants in South Africa, where the beasts can be very destructive to the limited park lands they live in, or transport them to East Africa, where the local elephants are endangered after years of poaching? Does this not risk diluting the genetic biodiversity of African elephants? Do we reintroduce predators such as wolves or sea otters, or do we let nature find its own way to deal with the blows we have dealt it? Do we care about snail darters?

Some intervention programs have had great success, such as the reintroduction of sea otters; others have not. Science is an ongoing process of learning, so we are bound to make mistakes such as the intentional introduction of multiflora rose for soil conservation or of European flies as "biocontrols" of gypsy moths. Moreover, reestablishing native populations with captive animals, or wild animals from another area, can lead to the spread of disease, as happened with orangutans.

Part of the problem is that ecologists who are trained as scientists have had to take on the role of engineers, which is a quite distinct endeavor. What's the difference? Science is a process of discovering and analyzing how the natural world works, whereas engineers create new ways to make things work. It is research versus design. I've often heard people say that if scientists can get us to the moon, then they can resolve the problems here on earth. But it was not scientists who sent spacecraft to the moon; it was engineers. The engineers used a lot of data gathered by astronomers, physicists, and other scientists, to be sure, but the trip to the moon was an engineering feat.

The problem with interventionist conservation, unlike the trip to the moon, is that science does not have sufficient data for the engineers to work with. Even with some well-warranted scientific pride among ecologists, we've had only a little over a century of ecological research. Nature has had millions of years to evolve complexly functioning ecosystems. Nature too has had its setbacks, hence the previous mass extinctions, but with time (and plenty of it) the living planet has always encountered a way to rebound. But time is not on our side now, given the accelerating rate of extinction. From a scientific outlook on life, human ecosystem engineers would have to do better than nature. From a religious point of view, we would have to do better than God. Either perspective is quite daunting.

As a scientist I often like to find out how things work, and when something in my house stops working I go in with considerable curiosity (and too much hubris), take the malfunctioning unit apart, and try to fix it. Even when I play Mr. Fix-it with something as simple as one of my children's toys, I end up either partially fixing it (with spare parts left over) or throwing the whole contraption away when it becomes unrecoverable. I've learned to keep away from my car engine or electronic devices, for my attempts to tinker with these engineering marvels usually magnify the problem and result in greater expenses when I take such things to the experts for repair.

In the living world, nature (and the biodiversity in it) is the equivalent of the expert engineer. Nature can grow a few blades of grass between the cracks in my sidewalk, while I can't get it to grow any on a bare patch of my lawn. We can't hope to match nature's ways, though we do the best we can. But like my auto mechanic, who never seems to be in a rush, nature can take a lot of time. Meanwhile we are inconvenienced, and eventually we will all pay dearly for losing its services. But what price should we pay now to help keep them? The price would be leaving nature plenty of room to carry on its business. We must give nature more than just a few fragmented patches. It is unseemly to stand over nature's shoulder giving unwanted and ill-informed advice (as my neighbors sometimes do to me). We must stand back—way back—and let the work be done. Preservation of expansive lands is the only way to do that, and the only way conservationists can ensure the perpetuation of their efforts is if we halt our population growth, or perhaps even reduce our numbers. Population "control" is the price we have to pay for nature's services.

Trees as Weeds

I have no idea how many poems have been written about trees, or how many works of art feature them. Fortunately, humans find them beautiful, just as ecologists find them valuable for the maintenance of biodiversity. It is as if there were something innate in the human mind that draws us toward nature. The term for such a concept, biophilia, was popularized by sociobiologist E. O. Wilson, who sees a little bit of bunny hugging in all of us. Wilson defined biophilia as "the connections that human beings subconsciously seek with the rest of life." In other words, our attraction to trees and nature is hardwired in our brains. If only it were so simple, then conservationists would not have to fight such an uphill battle.

Just as individuals change their outlook when life experiences bring them to new understandings or misunderstandings about the world around them, human cultures and ideals change through time. For example, during my grandmother's generation it was not unusual to get married and begin a family while still in one's teens, as she and her fiancé did. In many cases betrothal was an expectation of teenagers and a cause for great joy. Now our culture views early marriage as irresponsible and teen pregnancies as a social ill. When newspapers reported that the frequency of teen pregnancies went down in the United States, it was implicit that this was a good thing for the country.

We know we have an innate drive for reproduction, yet our cultural norms are strong enough to influence the stage of life when we reproduce. It is unknown whether or not we each have inherited something in our brain that necessarily leads to biophilia. Even if it were true, then our cultural constructs might be strong enough to overshadow our innate drive, just as with reproduction. Biophilia can quickly turn to its opposite, biophobia.

There is no reason why people could not see trees as weeds, nothing but an unwanted plant. After all, trees spring up in untended fields and grow into monstrous obstructions. The scraggly things block our view of the lands we have conquered, and take great effort to cut down. A nice brick building could provide just as much shade, and would be a lot tidier. In the fall, those ugly brown leaves drop everywhere, leaving us with the chore of getting rid of them. Leaves and pine needles that fall to the ground are really just tree excretions, for which one should have no more fondness than the droppings left by a neighbor's dog.

Such a view of trees may sound bizarre to you — at least I hope it does — but in some ways we do treat trees as weeds. I commented earlier on the first European settlers of the Ohio Valley needing to cut down trees to establish their farms. The trees were just in the way. Today it is no different. Forests are being destroyed at a rate of 14 million hectares per year,[5] and with them the biodiversity they harbor, in order to expand agricultural lands, housing, and commerce. As long as the human population keeps growing, there will be a need for more agricultural lands. Biophilia is no match for our innate drive to eat.

This presents conservationists with their greatest challenge. Lands set aside for nature to conduct its services often have to give way to "progress" as we search for food and energy. Even if human population growth were to cease tomorrow, such lands would eventually be broached, for we would still be consumers of the land. As the soil on existing farms erodes away, and the fossil fuels are depleted, new sources of nutrients and energy must be sought. The forests and the prairies hold the resources that could *support* our population. As rivers and lakes become unclean from our persistent use of them as garbage disposal systems, new waters will be tapped. It is the only way to *sustain* our current population — but it is no way to sustain a healthy planet full of biodiversity at every level.

If we can instill a global understanding of the values of biodiversity through education in our varied cultures, reserved lands may stand a chance, and future generations may reap the benefits of other species' actions. But I suspect it is more important, and perhaps easier, to halt the growth of our population. I would go as far as to advocate a drop in our numbers, given the devastation already wrought by six billion people.

Counting Future People

In 1972 The Club of Rome, an informal international organization, sponsored a study called the "Predicament of Mankind," which eventually got published as the classic book *The Limits to Growth*.[6] The idea was to use the best data and modeling techniques available to project future trends in population growth and its consequences. In particular, the authors were concerned with population, agricultural production, natural resources, industrial production, and pollution. Biodiversity loss had yet to feature in such studies. But along with similar research and publications, *The Limits to Growth* heightened awareness of population pressures and potential ecological disasters. This knowledge led to concerted efforts advancing agricultural production and pollution control, as well as family planning.

Twenty-five years later, I was surprised to read newspaper editorials giving snide postmortems on the projections made in the 1972 model. The crises that had been predicted in energy and food shortages were not as severe, and the population growth rate had slowed enough that we reached six billion people rather than the seven billion that had been projected. The gist of the editorials was that the authors of the project were wrong, mere fear mongers rather than competent scientists. I beg to differ. Part of the problem with their projections was that the data and modeling techniques were in a young stage of development. But the success of the book is that it did what it was meant to do: spur a sluggish human race into action by pointing out the big picture. And that picture is as grim as it ever was, because our population is still growing and a large portion of it is acting as if the earth had infinite resources. There is no safety in our numbers. Perhaps some more fear mongering wouldn't be such a bad thing.

Today our knowledge of earth's resources and our ability to project trends into the future has improved. Such projections will never be exact, for they are dealing with highly complex problems. Moreover, the models depend to a large portion on estimating how people will act during the course of their lives, and what innovations will become available in fields such as energy and agriculture that might improve the planet's ability to support us. So how accurate can they be? Already we have seen two projections of the future of the human population that are vastly different: the United Nations predicts that the human population will reach its peak at eleven billion in 2200, whereas the best estimate of an Austrian team put the number at nine billion in 2070.[7] That's a difference of two billion people and 130 years. A lot can happen to confound all predictions in that amount of time, with so many people.

The projections of human population growth depend, for example, on a continuing drop in fertility rates. Why should fertility rates decline? The *theory* is the demographic transition (chapter 5). It is assumed that more and more people will want to have fewer children, as has happened in many Western countries. How do we know that? We don't. Frankly, I think this assumption lulls us into unjustified complacency. There is no reason why the trend cannot reverse itself, as it did in the United States during the baby boom. Or teen pregnancies could become the "in" thing, as they were not so long ago. Without proactive efforts on the part of all peoples, governments, and religions, the worldwide decline in fertility simply will not happen. It takes education, particularly about something known as "family planning."

All in the Family

Our population continues to grow at an unprecedented pace because, in part, there are already so many people. This is the nature of exponential growth. But it is not just that there are too many people. It is that there are too many copulating people who don't take adequate precautions to prevent pregnancies. It need not be that way.

The need for family planning is clear: it is the most effective conservation tactic we have. Too often the simple process of establishing family planning clinics gets muddled in debates over abortion, or in some cases even contraceptive use is controversial. But such clinics can provide the necessary tools and educational resources, for family planning at its best is simply a matter of education.

Many people would like to stop at two children, not for altruistic reasons of saving the planet but for their own personal reasons. Yet they do not have adequate knowledge or resources to prevent pregnancies. The irony of our population problem is that many births are "unwanted" by the parents. As a species we have to find a way to get equitable family planning resources to everybody who wants them, and an education for all. If one has religious objections to contraception, then abstinence can be taught. If abortion is a problem, then contraceptive devices can prevent that.

There are many ways to avoid pregnancies. I once heard an African man say that using a condom is worthless because it is like eating candy with the wrapper still on. In that case there is always a vasectomy. A quick and safe surgical snip of the vas deferens, which carries the sperm, and there need not be any worry about unexpectedly expecting. The procedure is not entirely pain-less, I can assure you. But one day of minor pain gives a lifetime of freedom. In Iran the government set up a program sponsoring vasectomies for married men. A little education to allay their fears, free surgery, and the men were happy. "Now we can have sex whenever we want," said one. It has worked in Iran: in just a decade, the vasectomies, combined with government-provided contraceptives, cut the population growth rate in half.

Still, many parts of the world do not have the educational resources to get the word out, or the financial resources for family planning clinics or gov-ernmental programs. Moreover, many people who do have access to such re-sources still do not understand the big picture: how having many children stresses our resources and jeopardizes the sustainability of the planet. And people cannot solve problems of which they are unaware. Without a con-certed effort to change people's blissful ignorance, there will be no comple-tion of the demographic transition.

Pro-choice

In some countries that are experiencing drops in the fertility rate, a big part of the change has come from the empowerment of women. It is not just that they are getting into the workforce—that is simply employment. *Empower-ment* means that the women have more of a say in all aspects of their life, in-cluding their family life. They can choose whether to make the family larger or to stop at two children. I daresay that the male should take part in the choice as well. But a cultural milieu that nurtures careful and fair decision making in the family can go a long way toward stemming the growth rate of our population.[8]

Another cultural aspect of population growth has to do with society's ex-pectations. From the time a child is born, it is anticipated that he or she will grow up, get married, and have kids. More and more, couples choose not to have any children—and are looked upon as social mutants, particularly by the more conservative members of the community. Societal tolerance of a couple's right to choose to have no children would also help make this a more sustainable planet.

But there is a downside to empowerment and social freedom of choice. We cannot take away someone's right to have many children. Just as in the good ole days, big families can be fun. And that brings us back to Garrett

Hardin's tragedy of the commons: freedom in the commons brings ruin to all. Can we afford reproductive freedom, or should we impose a limit? A limit of one child per couple was imposed by China, and since abandoned. Dare we repeat the mistake? These are some of the tough questions that the current generation must answer.

Hardin's solution was "mutual coercion, mutually agreed upon."[9] That's a nice sentiment, but difficult to achieve. On the other hand, with an eye toward a sustainable future, we can craft subtle coercion. For example, there could be tax breaks for those who have fewer children, just as we pay farmers to leave their fields fallow. Right now in the United States, the tax breaks go to those who *do* have children. Which would be more fair? Society can decide fairness only when all the information is readily available, which is why those who are concerned must mobilize now to have an educational impact — just as did *The Limits to Growth*.

Beyond that the job of a scientist is done. The rest is up to the engineers of society: governments and the people who elect them, clergy, industry leaders, restoration ecologists, and environmental action programs such as The Nature Conservancy. That is one of the organizations to which I give some of my extra dollars, for the money goes directly to the threatened ecosystems.[10] However it gets done, the doing starts with you. I won't give you any more mantras; you know them.

The Specter of Population Decline

How many people can the earth *sustain*? It is a difficult question to answer. Ken Smail puts the number at two to three billion. Smail is a professor of anthropology at Kenyon College, here in Ohio. He took a bold step in suggesting that we need to go beyond a mere cessation of growth and work on population decline. In his words: "Within the next half-century, it will be essential for the human species to have in place a fully operational, flexibly designed, essentially voluntary, broadly equitable, and internationally coordinated set of initiatives focused on dramatically reducing the then-current world population by at least two-thirds to three-fourths."[11] Dramatic indeed!

Smail stops short of saying the words "population control," except to note that current human population growth is "out of control." But population reduction? It is the kind of term that puts economists and politicians in fits of denial, for their systems of creating wealth and building power bases depend on sustained growth. They talk about the "specter" of population decline in countries where the birthrate has dropped below the mortality rate, as if fewer people would spell doom and gloom. I'll be blunt: it is ridiculous to think for a moment that it would be more difficult to adjust our economic and social systems than to adapt our behavior to the incredible pressure we would be putting on our natural resources should we allow our population to continue to spin out of control. The problem now is that we have no precedent to work from, so we must be creative in how we remake our cultural world. But humans, being the wily and adaptable creatures they've always been, can certainly find a way.

Where does Smail get the two to three billion figure? He gets it from the 1950s, when world economic systems worked well for many, and the impact of our numbers was less. (The impact of our behavior is another issue.) He admits that the goal is not contrived by analysis of hard data, but reasons that we should shoot for it anyway. The juggernaut of population growth already has so much momentum that aiming for population reduction would help accelerate efforts to quell our current reproductive habits. I tend to agree: better to err on the side of prudence. We could always repopulate the world fairly quickly if need be, as we have in the recent past.

Smail makes another interesting point: "Population stabilization and subsequent reduction is the primary issue facing humanity; all other matters are subordinate."[12] A bold line like that is sure to bring out the critics, and it has. Tim Dyson, professor of population studies at the London School of Economics, was one: "A very strong argument can be made that changing consumption patterns is a more important issue facing humanity than limiting population growth (though this is not to deny that the latter is also a significant issue)."[13] Therein lies the heart of an ongoing debate. Is it how many of us there are, or is it how we live our lives that matters to the environment? No doubt we should be less wasteful, even if there were only three billion of us — simple conservation did not exist in the 1950s to the degree it should have. But Smail's point is this: it takes resources to sustain a population with a decent standard of living. That was Joel Cohen's point as well when he tried to answer the question of how many people the earth can support: it depends on how they want to live.[14] Even if we all become "vegetarian saints"[15] and live by the mantras expressed earlier in this chapter, our numbers matter.

Population numbers have always mattered with regard to our impact on other species. We saw the subtle effects of population growth when *Homo erectus* began spreading across Africa and beyond. The impact on other species became more clear with the hunting practices of *Homo sapiens*, whether on the moas of New Zealand or the mammoths and mastodons of North America. The human wedge became even more profound with the advent of agriculture some ten thousand years ago, and has increased exponentially since. Now we are to the point where we have to do something about our numbers, lest the biodiversity crisis become even worse. All other matters are indeed subordinate.

People Count

An understanding of the biodiversity crisis is critical to raising public awareness of the population problem, and motivating people to curb their reproductive habits. Public resistance and resentment can be abated with a greater understanding of the true issues at stake. Too often side issues get in the way. For example, environmentalists and conservationists are often accused of favoring wildlife over people. But it is not just a matter of saving monkeys versus saving people — that is not the choice.

One can see where the antienvironmentalist notions come from. After all, in the course of writing this book I've compared humans to weeds and the growth of our population to the spread of cancer cells, while repeatedly touting my fascination with baboons. That is hardly the kind of imagery to effectively assure a skeptic that I preferred humans to monkeys. But I chose a career in anthropology due to my love of people and fascination with what they can do. That does not eliminate my responsibility to exhibit my distress with what they sometimes have done.

My concern, and the concern of everyone, is making the world more sustainable. We cannot enjoy the services of millions of species if they are extinct. And they cannot persist if more and more of us invade their lands. Six billion people or more can't live with twelve million species — and we can't live without 'em.

We have taken advantage of our freedom in the commons of nature, and now we must act before it brings ruin to all. The alternatives are so clear: a better life for fewer people, or greater strife for more people. The solutions are so simple: conservation of the living world, and taking responsibility for our reproductive habits.

Back at the Office

In order to finish this book, I went back to my "office" on the Olentangy River to think. I wanted to end on a positive note, for as I told you I am an optimist, and I believe that the most adaptable species on earth can find a way to resolve the issues that threaten our future. So with the future in mind, I took my two young boys to the office with me.

At the edge of the river we didn't talk much, and they were too young for intellectual discussion about the relationship between human population growth and the biodiversity crisis. Instead we skipped stones across the water, and watched a caterpillar eat leaves. They too marveled at the dance of the damselflies. I wondered if their fascination with bugs was truly an innate biophilia, or if their parents had just instilled a curiosity about nature. As long as it was there, the source didn't matter. I saw that the future could be in good hands. They didn't want to control nature; they just wanted to appreciate it, enjoy it, and reap its practical benefits.

I hope that someday their children can do the same — skip stones on clean water and feast their eyes on a wide variety of bugs. But it is not my choice whether or not I get grandchildren. It is up to their generation, when it is their turn, to decide how to populate the earth and how to treat its resources. It is up to our generation to teach them well, and to set a good example.

Yes, the human population should continue to reproduce, but in a responsible fashion. It is the greatest gift we can give future generations — their lives, *and* a sustainable world for them to live in. But it is not just for future generations that we must be concerned. Without immediate intervention, we will all see a burgeoning human population and dramatic losses of biodiversity in our own lifetimes. How do I know? Because it is happening now. And so it is now that we must act. It is our turn, and we have to do this for ourselves.

As we grapple with our priorities, many species around the world are struggling to keep a foothold on life. From the gorillas of Africa to the pandas of China, animals and the ecosystems that support them falter. Miss Waldron's red colobus, passenger pigeons, and countless less conspicuous species are already gone. The remaining life-forms, be they mammals, birds, fish, or microbes in the soil, have increasingly greater difficulty finding some unfettered space and energy to have a life of their own. There is sorrow in this view of life, with its severed powers. It need not be. Nature's grandeur awaits a responsible human population.

Every good parent knows that what we do now for our children, they return manifold in due course. Likewise, every citizen of planet Earth should come to understand that what we do for biodiversity will come back to us as the free natural services that keep our world alive. We just need to spare nature the resources it needs.

Notes

1. SPARING NATURE

1. J. E. Cohen, *How Many People Can the Earth Support?* (New York: W. W. Norton & Co., 1995).
2. *Newsweek*, Feb. 15, 1999, p. 76.
3. The term "ecosystem" was coined by Tansley. A. G. Tansley, "The Use and Abuse of Vegetational Concepts and Terms," *Ecology* 42 (1935), 237–245.
4. Data kindly provided by naturalist John Watts of the Columbus Metropolitan Parks.
5. Estimates from the U.S. Census Bureau.
6. C. F. Westoff, "Population Growth: Large Problem, Low Visibility," *Politics and the Life Sciences* 16 (1997), 227.
7. S. Krech, *The Ecological Indian: Myth and History* (New York: W. W. Norton & Co., 1999).
8. L. Bromfield, *Malabar Farm* (London: Ballantine Books, 1947), 239.
9. C. S. Elton, *The Ecology of Invasions by Animals and Plants* (London: Methuen, 1958), 143.
10. As quoted in the *Columbus Dispatch*, Sept. 1, 2000, p. B6.
11. J. W. Bews, "The Ecological Viewpoint," *South African Journal of Science* 28 (1931), 11.
12. Mosquito larvae are an important food source for fish, and the adults are eaten by birds and bats, having transformed the nutrients of plants into concentrations of "meat" for the predators.

2. THE SCATTERED SEEDS

1. There are chemical signatures of the existence of life going back to 3.8 million years, but no fossils.
2. The "big five" mass extinctions are interspersed with other rises in extinction rates, though we are just coming to grips with their relative magnitudes and distributions. See A. B. Smith, A. S. Gale, and N.E.A. Monks, "Sea-Level Change and Rock-Record Bias in the Cretaceous: A Problem for Extinction and Biodiversity Studies," *Paleobiology* 27 (2001), 241–253. Data for the figure come from W. I. Ausich and N. G. Lane, *Life of the Past*, fourth edition (Upper Saddle River, N.J.: Prentice Hall, 1999).
3. Because humans are more similar genetically to chimps than to gorillas, yet both chimps and gorillas knuckle walk whereas our ancestors walked upright, it may be that knuckle walking evolved twice. If that were the case, it would be an example of parallel evolution.
4. A. R. Templeton, "Phylogenetic Inference from Restriction Endonuclease Cleavage Site Maps with Particular Reference to the Evolution of Man and the Apes," *Evolution* 37 (1983), 221–244.

5. The phrase "red in tooth and claw," quoted from Tennyson's poem (written before Darwin's *On the Origin of Species*), was often used in early critiques of "Darwinism" as being too violent and morally repugnant.

6. This study was conducted by William D'Arcy, but apparently never got published. The details from D'Arcy's lectures were given to me in lectures by Robert Sussman at Washington University.

7. P. R. Ehrlich and P. H. Raven, "Butterflies and Plants: A Study in Coevolution," *Evolution* 18 (1964), 586–608.

8. C. R. Peters and B. Maguire, "Wild Plant Foods of the Makapansgat Area: A Modern Ecosystems Analogue for *Australopithecus africanus* Adaptations," *Journal of Human Evolution* (1981), 565–583.

9. The monkeys were *Callithrix manicorensis* and *Callithrix acariensis*, two species of marmosets.

10. This is changing with greater focus on "bioinformatics," aided by the Internet. F. A. Bisby, "The Quiet Revolution: Biodiversity Informatics and the Internet," *Science* 289 (2000), 2309–2312.

11. E. O. Wilson, *The Diversity of Life* (London: Penguin Press, 1992).

12. P. H. Raven, "Disappearing Species: A Global Tragedy," *Futurist* 19 (1985), 8–14; N. E. Stork, "Measuring Global Biodiversity and Its Decline," in M. L. Reaka-Kudla, D. E. Wilson, and E. O. Wilson, eds., *Biodiversity II: Understanding and Protecting Our Biological Resources* (Washington, D.C.: Joseph Henry Press, 1997), 41–68.

13. N. E. Stork and K. G. Gaston, "Counting Species One by One," *New Scientist* 1729 (1990), 43–47; and Stork, "Measuring Global Biodiversity."

14. T. L. Erwin, "Tropical Forests: Their Richness in Coleoptera and Other Arthropod Species," *Coleoptera Bulletin* 36 (1982), 74–75; and Stork, "Measuring Global Biodiversity."

15. The subspecies name of Miss Waldron's red colobus is *Procolobus badius waldroni*.

16. J. F. Oates, M. Abedi-Lartey, W. S. McGraw, T. T. Struhsaker, and G. H. Whitesides, "Extinction of a West African Red Colobus Monkey," *Conservation Biology* 14 (2000), 1530.

17. R.D.E MacPhee and C. Flemming, "Requiem Æternam — the Last Five Hundred Years of Mammalian Species Extinctions," in R.D.E. MacPhee, ed., *Extinctions in Near Time: Causes, Contexts, and Consequences* (New York: Kluwer Academic/Plenum Publishers, 1999), 333–371.

18. M. E. Soulé, "What Do We Really Know about Extinction?" in C. M. Shonewald-Cox, S. M. Chambers, B. MacBryde, and W. L. Thomas, eds., *Genetics and Conservation* (London: Benjamin/Cummings Publishing, 1983), 112.

19. R. W. Martin, "Biological Diversity: Divergent Views on Its Status and Diverging Approaches to Its Conservation," in R. Bailey, ed., *Earth Report 2000* (New York: McGraw-Hill, 2000), 235.

20. P. R. Ehrlich and A. H. Ehrlich, *Betrayal of Science and Reason: How Anti-Environmental Rhetoric Threatens Our Future* (Washington, D.C.: Island Press, 1996), 113.

21. Excepting identical twins or propagated cuttings, of course.

22. W. Berry, *Life Is a Miracle: An Essay against Modern Superstition* (Washington, D.C.: Counterpoint, 2000), 7.

23. I should note that cloning animals may have advantages for medical research, just not for the production of farm animals.

24. For a more thorough explanation, see M. L. Rosenzweig, *Species Diversity in Space and Time* (Cambridge: Cambridge University Press, 1995).

25. A. Hendry, J. Wenburg, P. Bentezen, E. Volk, and T. Quinn, "Rapid Evolution of Reproductive Isolation in the Wild: Evidence from Introduced Salmon," *Science* 290 (2000), 516–518.

26. C. Darwin, *On the Origin of Species by Means of Natural Selection* (London: John Murray, 1859), 67.

3. THE HUMAN WEDGE

1. H. McHenry, "Early Hominid Stature," *American Journal of Physical Anthropology* 85 (1991), 149–158.
2. From Stevenson's poem "Nest Eggs."
3. C. B. Ruff and A. Walker, "Body Size and Body Shape," in A. Walker and R. Leakey, eds., *The Nariokotome* Homo erectus *Skeleton* (Cambridge, Mass.: Harvard University Press, 1993).
4. A. K. Behrensmeyer, N. E. Todd, R. Potts, and G. E. McBrinn, "Late Pliocene Faunal Turnover in the Turkana Basin, Kenya and Ethiopia," *Science* 278 (1997), 1589–1594.
5. J. K. McKee, "Turnover Patterns and Species Longevity of Large Mammals from the Late Pliocene and Pleistocene of Southern Africa: A Comparison of Simulated and Empirical Data," *Journal of Theoretical Biology* 172 (1995), 141–147; J. K. McKee, "Faunal Turnover Rates and Mammalian Biodiversity of the Late Pliocene and Pleistocene of Eastern Africa," *Paleobiology* 27 (2001), 500–511.
6. R. S. Devine, *Alien Invasion: America's Battle with Non-native Animals and Plants* (Washington, D.C.: National Geographic Society, 1998).
7. L. Gabunia, A. Vekua, D. Lordkipanidze, C. Swisher, R. Ferring, A. Justus, M. Nioradze, M. Tvalchrelidae, S. C. Anton, G. Bosinski, O. Jöris, M. A. de Lumley, G. Majsuradze, and A. Mouskhelishvili, "Earliest Pleistocene Hominid Cranial Remains from Dmanisi, Republic of Georgia: Taxonomy, Geological Setting, and Age," *Science* 288 (2000), 1019–1025.
8. R. Byrne and J. Byrne, "Leopard Killers of Mahale," in R. L. Ciochon and R. Nisbett, eds., *The Primate Anthology: Essays on Primate Behavior, Ecology, and Conservation from Natural History* (Upper Saddle River, N.J.: Prentice Hall, 1997), 113–118.
9. R. G. Klein, "Human Evolution and Large Mammal Extinctions," in E. S. Vrba and G. W. Schaller, eds., *Antelopes, Deer, and Relatives, Present and Future: Fossil Record, Behavioral Ecology, Systematics, and Conservation* (New Haven: Yale University Press, 2000), 128–139. Note that the figures I use are adjusted from his data to account for the variable length of the time periods he uses.
10. Because of the mode of analysis, and the cutoff dates Klein chose, the highest number of extinctions appears to be before the human entry. But keep in mind that we are looking at the number of extinct genera as a proportion of all genera, not extinction rates. We do not know when they went extinct, just that they are now extinct and were lost to the fossil record around that time. Some of my own research deals with these issues (e.g., see the references in note 5 to this chapter).
11. This is the topic of my last book: J. K. McKee, *The Riddled Chain: Chance, Coincidence, and Chaos in Human Evolution* (New Brunswick, N.J.: Rutgers University Press, 2000).
12. The nature of the cave's occupation and the exact use of fire is disputed, but such details should not detract from the evident lifestyle of *Homo erectus*.
13. N. T. Boaz and R. L. Ciochon, "The Scavenging of 'Peking Man,'" *Natural History* 110 (2001), 46–51.
14. N. Owen-Smith, "The Interaction of Humans, Megaherbivores, and Habitats in the Late Pleistocene Extinction Event," in R.D.E. MacPhee, ed., *Extinctions in Near Time: Causes, Contexts, and Consequences* (New York: Kluwer Academic/Plenum Publishers, 1999), 57–69.
15. We do, however, have evidence of human occupation of Northeast Asia by 1.36 million years ago and of European arctic regions as early as 35,000–40,000 years ago. R. X. Zhu,

K. A. Hoffman, R. Potts, C. L. Deng, Y. X. Pan, B. Guo, C. D. Shi, Z. T. Guo, B. Y. Yuan, Y. M. Hou, and W. W. Huan, "Early Presence of Humans in Northeast Asia," *Nature* 413 (2001), 413–417; P. Pavlov, J. I. Svendsen, and S. Indrelid, "Human Presence in the European Arctic Nearly 40,000 Years Ago," *Nature* 413 (2001), 64–67.

16. P. S. Martin and D. W. Steadman, "Prehistoric Extinctions on Islands and Continents," in R.D.E. MacPhee, ed., *Extinctions in Near Time: Causes, Contexts, and Consequences* (New York: Kluwer Academic/Plenum Publishers, 1999), 17–55.

17. N. Owen-Smith, "Pleistocene Extinctions: The Pivotal Role of Megaherbivores," *Paleobiology* 13 (1987), 351–362.

18. Owen-Smith, "Pleistocene Extinctions."

19. J. Alroy, "A Multispecies Overkill Simulation of the End-Pleistocene Megafaunal Mass Extinction," *Science* 292 (2001), 1893–1896.

20. I should note that Alroy used higher rates of population growth than I did earlier in this chapter. However, if one adjusts back the time of initial human entry from the time used in Alroy's model, then the eventual consequences are the same.

21. Alroy, "Multispecies Overkill Simulation," 1896.

22. R. G. Roberts, T. F. Flannery, L. K. Ayliffe, H. Yoshida, J. M. Olley, G. J. Prideaux, G. M. Laslett, A. Baynes, M. A. Smith, R. Jones, and B. L. Smith, "New Ages for the Last Australian Megafauna: Continent-wide Extinction about 46,000 Years Ago," *Science* 292 (2001), 1888–1892.

23. L. Dayton, "Mass Extinctions Pinned on Ice Age Hunters," *Science* 292 (2001), 1819.

24. Roberts et al., "New Ages."

25. R. N. Holdaway and C. Jacomb, "Rapid Extinction of the Moas (Aves: Dinornithiformes): Model, Test, and Implications," *Science* 287 (2000), 2250–2254.

26. J. Diamond, "Blitzkrieg against the Moas," *Science* 287 (2000), 2170–2171.

27. Owen-Smith, "Pleistocene Extinctions," 65.

4. GENESIS OF A CRISIS

1. M. C. Stiner, N. D. Munro, and T. A. Surovell, "The Tortoise and the Hare: Small-Game Use, the Broad-Spectrum Revolution, and Paleolithic Demography," *Current Anthropology* 41 (2000), 39–58.

2. V. G. Childe, *Social Evolution* (London: Watts, 2000).

3. P. J. Richerson, R. Boyd, and R. L. Bettinger, "Was Agriculture Impossible during the Pleistocene but Mandatory during the Holocene? A Climate Change Hypothesis," *American Antiquity* 66 (2001), 387–411.

4. R. J. Braidwood, *Prehistoric Men* (Chicago: Chicago Natural History Museum, 1951); R. J. Braidwood and G. R. Willey, "Conclusions and Afterthoughts," in R. J. Braidwood and G. R. Willey, eds., *Courses towards Urban Life* (Chicago: Aldine, 1962), 330–359. Incidentally, the character Professor Ravenwood of the Indiana Jones movie was modeled after Robert Braidwood.

5. This "orthogenetic" point of view, however, has recently been revived by Robert Wright, *Nonzero: The Logic of Human Destiny* (New York: Pantheon Books, 2000).

6. D. Rindos, *The Origins of Agriculture: An Evolutionary Perspective* (San Diego: Academic Press, 1984).

7. B. D. Smith, *The Emergence of Agriculture* (New York: Scientific American Library, 1995).

8. M. Rosenberg, "Cheating at Musical Chairs: Territoriality and Sedentism in an Evolutionary Context," *Current Anthropology* 39 (1998), 657.

9. Rosenberg, "Cheating at Musical Chairs," 660.

10. The egg came first in evolution. The egg came from a bird that was almost a chicken.

11. It should be noted that one can play with this model, giving it higher growth rates more recently and slower or variable growth rates in the past, but almost invariably the exponential upswing around ten thousand years ago becomes evident.

12. J. Diamond, *Guns, Germs, and Steel: The Fates of Human Societies* (New York: W. W. Norton & Co., 1997); see also McKee, *Riddled Chain*, for more on autocatalysis.

13. Haeckel is often credited with coining the word "ecology," but apparently it should be attributed to Hanns Reiter, who first used it in 1885. Haeckel *did* bring the term into common use.

14. J. W. Bennett, *The Ecological Transition: Cultural Anthropology and Human Adaptation* (New York: Pergamon Press, 1976), 3.

15. Rindos, *Origins of Agriculture*, 6.

16. Bennett, *Ecological Transition*, 5.

17. M. J. and J. M. Anderson, "Biodiversity and Ecosystem Function in Agricultural Systems," in E.-D. Schulze and H. A. Mooney, eds., *Biodiversity and Ecosystem Function* (Berlin: Springer-Verlag, 1993), 15–41.

18. E. Pennisi, "The Push to Pit Genomics against Fungal Pathogens," *Science* 292 (2001), 2273–2274.

19. A. S. Moffat, "Finding New Ways to Fight Plant Disease," *Science* 292 (2001), 2270–2273.

20. W. H. Drury, Jr., *Chance and Change: Ecology for Conservationists* (Berkeley: University of California Press, 1998), 190.

21. *Webster's New Collegiate Dictionary* (Springfield, Mass.: G. & C. Merriam Co., 1975), my trusty companion since I graduated from high school.

22. Bews, "Ecological Viewpoint," 12–13.

23. C. L. Redman, *Human Impact on Ancient Environments* (Tucson: University of Arizona Press, 1999).

24. P. B. deMenocal, "Cultural Responses to Climate Change during the Late Holocene," *Science* 292 (2001), 667–673.

25. deMenocal, "Cultural Responses."

26. D. A. Hodell, M. Brenner, J. H. Curtis, and T. Guilderson, "Solar Forcing of Drought Frequency in the Maya Lowlands," *Science* 292 (2001), 1367–1371.

27. E. H. Roseboom and F. P. Weisenburger, *A History of Ohio* (Columbus: Ohio Historical Society, 1973).

28. Diamond (*Guns, Germs, and Steel*) presents a superb and thorough account of this story.

29. Krech, *Ecological Indian*, 99.

30. Krech, *Ecological Indian*.

31. Roseboom and Weisenburger, *History of Ohio*, 5.

32. Cohen, *How Many People*.

33. S. L. Pimm, *The World According to Pimm: A Scientist Audits the Earth* (New York: McGraw-Hill, 2001), building on the work of P. M. Vitousek, P. R. Ehrlich, A. H. Ehrlich, and P. A. Matson, "Human Appropriation of the Products of Photosynthesis," *Bioscience* 36 (1986), 368–373. Pimm's book makes for excellent reading on this subject.

34. D. Tilman, J. Fargione, B. Wolff, C. D'Antonio, A. Dobson, R. Howarth, D. Schindler, W. H. Schlesinger, D. Simberloff, and D. Swackhamer, "Forecasting Agriculturally Driven Global Environmental Change," *Science* 292 (2001), 281–284.

35. P. M. Vitousek, H. A. Mooney, J. Lubchenco, and J. M. Melillo, "Human Domination of Earth's Ecosystems," *Science* 277 (1997), 494–499.

36. Tilman et al., "Forecasting," 283.

37. P. R. Ehrlich, *Human Natures: Genes, Cultures, and the Human Prospect* (Washington, D.C.: Island Press, 2000), and comments from an October 24, 2001, WGN Chicago radio interview.

5. Germs of Existence

1. I last visited China in 1991. Apparently some birds have returned since that time.
2. J. Shapiro, *Mao's War against Nature: Politics and the Environment in Revolutionary China* (Cambridge: Cambridge University Press, 2001).
3. S. Tuljapurkar, N. Li, and C. Boe, "A Universal Pattern of Mortality Decline in the G7 Countries," *Nature* 405 (2000), 789–792.
4. The natality rate is also sometimes called the "crude fertility rate."
5. All the data presented in this section come from the U.S. Census Bureau.
6. The true replacement rate depends on the structure of the population and changes in mortality.
7. Although both the United States and China have a 0.9 percent growth rate, much of the U.S. rate is attributable to immigration.
8. South Africa's growth rate had negative values during much of the early 1990s, largely due to emigration.
9. Weasel, "Pop with Weasel," *The ZPG Reporter* 31 (1999), 1.
10. W. Lutz, W. Sanderson, and S. Scherbov, "The End of World Population Growth," *Nature* 412 (2001), 543–545.
11. J. R. Weeks, *Population: An Introduction to Concepts and Issues* (Belmont, Calif.: Wadsworth Publishing Co., 1978), 278.
12. Lutz et al., "End of World Population Growth," 544.
13. McKee, *Riddled Chain*. My apologies for the repetition to those who read this before, but the exercise is worth pursuing in greater detail here.
14. I put "conservative" in quotes because Limbaugh's ideas are actually quite radical.
15. P. R. Ehrlich and J. Holdren, "Impact of Population Growth," *Science* 171 (1971), 1212–1217.
16. Cohen, *How Many People*.
17. From A. A. Milne's Winnie-the-Pooh tales.
18. D. M. Raup, "The Role of Extinction in Evolution," *Proceedings of the National Academy of Science* 91 (1994), 6758.
19. R. Leakey and R. Lewin, *The Sixth Extinction: Patterns of Life and the Future of Humankind* (New York: Anchor Books, 1995).
20. J. D. Skinner and R.H.N. Smithers, *The Mammals of the Southern African Subregion* (Pretoria: University of Pretoria, 1990).
21. According to the IUCN Red List.
22. No bigger than Ireland, including Northern Ireland.
23. For a complete look at the status of biodiversity in the United States see B. A. Stein, L. S. Kutner, and J. S. Adams, eds., *Precious Heritage: The Status of Biodiversity in the United States* (Oxford: Oxford University Press, 2000).
24. Stork, "Measuring Global Biodiversity."
25. D. J. Forester and G. E. Machlis, "Modeling Human Factors That Affect the Loss of Biodiversity," *Conservation Biology* 10 (1996), 1253–1263.
26. http://www.census.gov/ipc/www/

27. These are the categories of the IUCN–World Conservation Union Red List of Threatened Mammals. Extinct and extinct in the wild are additional categories, but the numbers are highly variable depending on how well-known an area has been studied in the past. Moreover, we know that all places have a high number of extinct species, depending on how far back in time you go.

28. For confirmation from sub-Saharan Africa, see also A. Balmford, J. L. Moore, T. Brooks, N. Burgess, L. A. Hansen, P. Williams, and C. Rahbek, "Conservation Conflicts across Africa," *Science* 291 (2001), 2616–2619.

29. D. Fooce and J. K. McKee, "Human Population Size as a Predictor of Threatened Species," *American Journal of Physical Anthropology Supplement* 23 (2002), 71.

30. M. L. McKinney, "Role of Human Population Size in Raising Bird and Mammal Threat among Nations," *Animal Conservation* 4 (2001), 45–57.

31. J. K. McKee, P. W. Sciulli, C. D. Fooce, and T. A. Waite, "Predicting Biodiversity Threats Associated with Human Population Growth," in preparation.

32. In detail, the model is log-transformed and looks like this: log threatened species per $10^6 \text{ km}^2 = -1.534 + 0.691 \times$ log species richness $+ 0.259 \times$ human population density.

33. J. K. McKee et al., "Predicting Biodiversity Threats."

34. N. Myers, R. A. Mittermeier, C. G. Mittermeier, G.A.B. da Fonseca, and J. Kent, "Biodiversity Hotspots for Conservation Priorities," *Nature* 403 (2000), 853.

35. Meyers et al., "Biodiversity Hotspots."

36. R. P. Cincotta and R. Engelman, *Nature's Place: Human Population and the Future of Biological Diversity* (Washington, D.C.: Population Action International, 2000).

37. Meyers et al., "Biodiversity Hotspots."

38. *The Lorax* by Theodor Seuss Geisel.

39. J. Liu, M. Linderman, Z. Ouyang, L. An, J. Yang, and H. Zhang, "Ecological Degradation in Protected Areas: The Case of Wolong Nature Reserve for Giant Pandas," *Science* 292 (2001), 98–100. The point remains the same even though Wolong is atypical of China's panda reserves, according to C. J. Loucks, Z. Lü, E. Dinerstein, H. Wang, D. M. Olson, C. Zhu, and D. Wang, "Giant Pandas in a Changing Landscape," *Science* 294 (2001), 1465. These authors also emphasize the efforts needed to save China's 1,100 remaining wild pandas.

6. The Great Restrictive Law

1. Khakibos is *Bidens pilosa*, "bi-dens" referring to the two "teeth" of the black jacks. Its cousin, *Bidens formosa*, is commonly known as cosmos, a pretty flower—but one that is also invasive to South Africa.

2. P. D. Tyson, *Climatic Change and Variability in Southern Africa* (Cape Town: Oxford University Press, 1987).

3. Tilman et al., "Forecasting."

4. O. E. Sala, F. S. Chapin II, J. J. Armesto, E. Berlow, J. Bloomfield, R. Dirzo, E. Huber-Sanwald, L. F. Huenneke, R. B. Jackson, A. Kinzig, R. Leemans, D. M. Lodge, H. A. Mooney, M. Oesterheld, N. L. Poff, M. T. Sykes, B. H. Walker, M. Walker, and D. H. Wall, "Global Biodiversity Scenarios for the Year 2100," *Science* 287 (2000), 1770–1774.

5. D. J. Bender, T. A. Contreras, and L. Fahrig, "Habitat Loss and Population Decline: A Meta-analysis of the Patch Size Effect," *Ecology* 79 (1998), 517–533.

6. J. G. Fleagle, *Primate Adaptation and Evolution* (San Diego: Academic Press, 1999), 245.

7. Y. Baskin, "A Sickening Situation: Emerging Pathogens Pose a Threat to Wildlife," *Natural History* 109 (2000), 24–27.

8. Bender et al., "Habitat Loss."

9. C. Gascon, G. B. Williamson, and G.A.B. da Fonsecea, "Receding Forest Edges and Vanishing Reserves," *Science* 288 (2000), 1356–1358.

10. T. A. Gavin, P. W. Sherman, E. Yensen, and B. May, "Population Genetic Structure of the Northern Idaho Ground Squirrel (*Spermophilus brunneus brunneus*)," *Journal of Mammalogy* 80 (1999), 156–168.

11. Drury, *Chance and Change.*

12. T. Wilkinson, "Prometheus Unbound," *Nature Conservancy* 51 (2001), 12–20.

13. M. L. Rosenzweig, "Loss of Speciation Rate Will Impoverish Future Diversity," *Proceedings of the National Academy of Science* 98 (2001), 5404. For a complete analysis see also S. L. Pimm, *The Balance of Nature? Ecological Issues in the Conservation of Species and Communities* (Chicago: University of Chicago Press, 1991).

14. G. Cowlishaw, "Predicting the Pattern of Decline of African Primate Diversity: An Extinction Debt from Historical Deforestation," *Conservation Biology* 13 (1999), 1183–1193.

15. Soulé, "What Do We Really Know," 112.

16. D. Ferber, "Human Diseases Threaten Great Apes," *Science* 289 (2000), 1277–1278.

17. Harvard biologist Stephen Palumbi notes that we increase the evolutionary rate not only of diseases but also of invasive species that quickly adapt to new environments, fish that evolve to escape our nets, and more. S. R. Palumbi, "Humans as the World's Greatest Evolutionary Force," *Science* 293 (2001), 1786–1790.

18. R. M. Rolland, G. Hausfater, B. Marshall, and S. B. Levy, "Antibiotic-Resistant Bacteria in Wild Primates: Increased Prevalence in Baboons (*Papio cynocephalus*) Feeding on Human Refuse," *Applied and Environmental Microbiology* 49 (1985), 791–794. But see also E. Routman, R. D. Miller, J. Phillips-Conroy, and D. L. Hartl, "Antibiotic Resistance and Population Structure in *Escherichia coli* from Free-Ranging African Yellow Baboons (*Papio cynocephalus*)," *Applied and Environmental Microbiology* 50 (1985), 749–754, who found no such effect among a different baboon population.

19. M. Gilliver, M. Bennett, M. Begon, S. Hazel, and C. Hart, "Enterobacteria: Antibiotic Resistance Found in Wild Rodents," *Nature* 401 (1999), 233–234.

20. G. H. Boettner, J. S. Elkinton, and C. J. Boettner, "Effects of a Biological Control Introduction on Three Nontarget Native Species of Saturniid Moths," *Conservation Biology* 14 (2000), 1798–1806.

21. M. Enserink, "Biological Invaders Sweep In," *Science* 285 (1999), 1834–1836.

22. Enserink, "Biological Invaders."

23. J. Weiner, *The Beak of the Finch: Evolution in Real Time* (London: Vintage, 1995).

24. Information on the gray duck, spotted owl, and hartebeest/blesbok comes from J. M. Rhymer and D. Simberloff, "Extinction by Hybridization and Introgression," *Annual Review of Ecological Systems* 27 (1996), 83–109.

25. For a riveting discussion of the wolf problem, see A. Chase, *Playing God in Yellowstone: The Destruction of America's First National Park* (San Diego: Harcourt Brace, 1987).

26. For the trout, see H. A. Mooney and E. E. Cleland, "The Evolutionary Impact of Invasive Species," *Proceedings of the National Academy of Science* 98 (2001), 5446–5451. For homogenization, see F. J. Rahel, "Homogenization of Fish Faunas across the United States," *Science* 288 (2000), 854–856.

27. Vitousek et al., "Human Domination"; T. J. Crowley, "Causes of Climate Change of the Past 1000 Years," *Science* 290 (2000), 270–277; P. A. Stott, S.F.B. Tett, G. S. Jones, M. R. Allen, J.F.B. Mitchell, and G. J. Jenkins, "External Control of 20th Century Temperature by Natural and Anthropogenic Forcings," *Science* 290 (2000), 2133–2137; S. Levitus, J. I. Antonov, J. Wang, T. L. Delworth, K. W. Dixon, and A. J. Broccoli, "Anthropogenic Warming of the Earth's Climate System, *Science* 292 (2001), 267–270; T. P. Barnett, D. W.

Pierce, and R. Schnur, "Detection of Anthropogenic Climate Change in the World's Oceans," *Science* 292 (2001), 270–274.

28. F. Drake, *Global Warming: The Science of Climate Change* (London: Arnold, 2000).
29. Sala et al., "Global Biodiversity Scenarios."
30. K. Krajick, "Arctic Life, on Thin Ice," *Science* 291 (2001), 424–425.
31. E. S. Vrba, "Mammals as a Key to Evolutionary Theory," *Journal of Mammalogy* 73 (1992), 1–28.
32. Gelada baboons are also under threat because the expanding human population wants to exterminate them in order to control disease.
33. P. R. Epstein, "Is Global Warming Harmful to Health?" *Scientific American* 283 (2000), 50–57.
34. D. J. Rogers and S. E. Randolf, "The Global Spread of Malaria in a Future, Warmer World," *Science* 289 (2000), 1763–1766.
35. Krajick, "Arctic Life."
36. J. R. Etterson and R. G. Shaw, "Constraint to Adaptive Evolution in Response to Global Warming," *Science* 294 (2001), 151–154.
37. M. C. Rutherford, G. F. Midgley, W. J. Bond, L. W. Powrie, R. Roberts, and J. Allsopp, "Plant Biodiversity," in G. Kiker, ed., *Climate Change Impacts in Southern Africa: Report to the National Climate Change Committee* (Pretoria: Department of Environmental Affairs and Tourism, 2000).
38. S. L. LaDeau and J. S. Clark, "Rising CO_2 Levels and the Fecundity of Forest Trees," *Science* 292 (2001), 95–98.

7. Good to the Last Drop

1. P. H. Gleick, "Making Every Drop Count," *Scientific American* 284 (2001), 41.
2. P. S. Levin and M. H. Schiewe, "Preserving Salmon Biodiversity," *American Scientist* 89 (2001), 220–227.
3. R. A. Kerr, "West's Energy Woes Threaten Salmon Runs," *Science* 291 (2001), 1470–1471.
4. J.B.C. Jackson, M. X. Kirby, W. H. Berger, K. A. Bjorndal, L. W. Botsford, B. J. Bourque, R. H. Bradbury, R. Cooke, J. Erlandson, J. A. Estes, T. P. Hughes, S. Kidwell, C. B. Lange, H. S. Lenihan, J. M. Pandolfi, C. H. Peterson, R. S. Steneck, M. J. Tegner, and R. R. Warner, "Historical Overfishing and the Recent Collapse of Coastal Ecosystems," *Science* 293 (2001), 629–638.
5. S. McKinnell and A. J. Thomson, "Recent Events Concerning Atlantic Salmon Escapees in the Pacific," *Journal of Marine Science* 54 (1997), 1221–1225.
6. I was surprised to learn that the Cuyahoga River also burned in 1952!
7. Information on the BWWA comes from M. Heinselman, *The Boundary Waters Wilderness Ecosystem* (Minneapolis: University of Minnesota Press, 1996).
8. C. L. Dybas, "Aliens," *Wildlife Conservation* 104 (2001), 56–60.
9. Enserink, "Biological Invaders."
10. Jackson et al., "Historical Overfishing."
11. R. L. Naylor, R. J. Goldburg, J. H. Primavera, N. Kautsky, M.C.M. Beveridge, J. Clay, C. Folke, J. Lubchenco, H. A. Mooney, and M. Troell, "Effect of Aquaculture on World Fish Supplies," *Nature* 405 (2000), 1017–1024.
12. Naylor et al., "Effect of Aquaculture," 1018.
13. R. L. Naylor, S. L. Williams, and D. R. Strong, "Aquaculture: A Gateway for Exotic Species," *Science* 294 (2001), 1655–1656.

14. P. Colinvaux, *Why Big Fierce Animals Are Rare: An Ecologist's Perspective* (Princeton, N.J.: Princeton University Press, 1978), 96.

15. D. Suzuki, *The Sacred Balance: Rediscovering Our Place in Nature* (Vancouver: Greystone Books, 1997), 80.

16. A. Goudie, *The Human Impact on the Natural Environment*, third edition (Cambridge, Mass.: MIT Press, 1990).

17. G. Hardin, "The Tragedy of the Commons," *Science* 162 (1968), 1244.

18. D. Martindale, "Sweating the Small Stuff," *Scientific American* 284 (2001), 52-53.

19. Tilman et al., "Forecasting."

20. Cohen, *How Many People*.

21. C. Brain, "Water Gathering by Baboons in the Namib Desert," *South African Journal of Science* 84 (1988), 590-591.

22. Gleick, "Making Every Drop Count."

23. C. J. Vörösmarty, P. Green, J. Salisbury, and R. B. Lammers, "Global Water Resources: Vulnerability from Climate Change and Population Growth," *Science* 289 (2000), 284-288.

24. Vörösmarty et al., "Gobal Water Resources," 287.

8. BIODIVERSITY IN ACTION

1. For further explanation read McKee, *Riddled Chain*.

2. A. Moorehead, *Darwin and the Beagle* (New York: Harper & Row, 1969), 57.

3. Myers et al., "Biodiversity Hotspots."

4. S. L. Pimm, M. Ayres, A. Balmford, G. Branch, K. Brandon, T. Brooks, R. Bustamante, R. Constanza, R. Cowling, L. M. Curran, A. Dobson, S. Farber, G.A.B. da Fonseca, C. Gascon, R. Kitching, J. McNeely, T. Lovejoy, R. A. Mittermeier, N. Myers, J. A. Patz, B. Raffle, D. Rapport, P. Raven, C. Roberts, J. P. Rodriguez, A. B. Rylands, C. Tucker, C. Safina, C. Samper, M.L.J. Stiassny, J. Spriatna, D. H. Wall, and D. Wilcove, "Can We Defy Nature's End?" *Science* 293 (2001), 2207-2208.

5. N. Myers, "The World's Forests and Their Ecosystem Services," in G. C. Daily, ed., *Nature's Services: Societal Dependence on Natural Ecosystems* (Washington, D.C.: Island Press, 1997), 215-236; see also J. Q. Chambers, N. Higuchi, E. S. Tribuzy, and S. E. Trubore, "Carbon Sink for a Century," *Nature* 410 (2001), 429.

6. P. R. Ehrlich and A. H. Ehrlich, "The Value of Biodiversity," *Ambio* 21 (1992), 219-226.

7. N. Myers, "Biodiversity's Genetic Library," in G. C. Daily, ed., *Nature's Services: Societal Dependence on Natural Ecosystems* (Washington, D.C.: Island Press, 1997), 255-273.

8. I put "natural" in quotes, because all genes are natural, and agriculture has been "unnatural" (read human-dominated) since its origins ten thousand years ago.

9. O. E. Sala and J. M. Paruelo, "Ecosystem Services in Grasslands," in G. C. Daily, ed., *Nature's Services: Societal Dependence on Natural Ecosystems* (Washington, D.C.: Island Press, 1997), 237-252.

10. Y. Luo, S. Wan, D. Hui, and L. L. Wallace, "Acclimatization of Soil Respiration to Warming in a Tall Grass Prairie," *Nature* 413 (2001), 622-625.

11. Y. Baskin, *The Work of Nature: How the Diversity of Life Sustains Us* (Washington, D.C.: Island Press, 1997).

12. Sala and Paruelo, "Ecosystem Services in Grasslands," 247-248.

13. In 1992 dollars. D. Pimentel, C. Harvey, P. Resosudarmo, K. Sinclair, D. Kurz, M. McNair, S. Crist, L. Shpritz, L. Fitton, R. Saffouri, and R. Blair, "Environmental and Economic Costs of Soil Erosion and Conservation Benefits," *Science* 267 (1995), 1117-1123.

14. The words come from a letter allegedly written to President Franklin Pierce in 1855, but the letter has not resurfaced, and there is historical debate as to whether Chief Seattle wrote the words. That should not detract from the power of the words here.

15. C. H. Peterson and J. Lubchenco, "Marine Ecosystem Services," in G. C. Daily, ed., *Nature's Services: Societal Dependence on Natural Ecosystems* (Washington, D.C.: Island Press, 1997), 177–194; S. Postel and S. Carpenter, "Freshwater Ecosystem Services," in G. C. Daily, ed., *Nature's Services: Societal Dependence on Natural Ecosystems* (Washington, D.C.: Island Press, 1997), 195–214.

16. M. Fischetti, "Drowning New Orleans," *Scientific American* 284 (2001), 76–85.

17. R. G. Torricelli, *Quotations for Public Speakers: A Historical, Literary, and Political Anthology* (New Brunswick, N.J.: Rutgers University Press, 2001), 74.

18. R. M. Cowling, R. Costanza, and S. I. Higgins, "Services Supplied by South African Fynbos Ecosystems," in G. C. Daily, ed., *Nature's Services: Societal Dependence on Natural Ecosystems* (Washington, D.C.: Island Press, 1997), 345–362.

19. With all due respect to Paul and Anne Ehrlich, this is my remake of their "rivet hypothesis," in which they compare species in ecosystems to rivets holding on the wing of a plane. The plane might fly with some missing rivets, but one more loss could lead to a crash. It is just difficult to tell what may be the critical number of rivets . . . or species. P. R. Ehrlich and A. Ehrlich, *Extinction: The Causes and Consequences of the Disappearance of Species* (New York: Random House, 1981).

20. Actually Van Gogh cut off his pinna, which gathers sound, not the ear, which hears it. Like some body parts, and some species, it proved to be expendable.

21. J. H. Lawton and V. K. Brown, "Redundancy in Ecosystems," in E.-D. Schulze and H. A. Mooney, eds., *Biodiversity and Ecosystem Function* (Berlin: Springer-Verlag, 1993), 255–270.

22. The scientific names are *Saguinus fuscicollis* and *Saguinus imperator* respectively. See J. Terborgh, *Five New World Primates: A Study in Comparative Ecology* (Princeton, N.J.: Princeton University Press, 1983).

23. Heinselman, *Boundary Waters.*

24. Keystone species may be categorized into predators, herbivores, pathogens and parasites, competitors, mutualists, earth movers, and system processors, as by W. J. Bond, "Keystone Species," in E.-D. Schulze and H. A. Mooney, eds., *Biodiversity and Ecosystem Function* (Berlin: Springer-Verlag, 1993), 237–253.

25. R. T. Paine, "Food Web Complexity and Species Diversity," *American Naturalist* 100 (1966), 65–75; R. T. Paine, "The *Pisaster-Tegula* Interaction: Prey Patches, Predator Food Preference, and Intertidal Community Structure," *Ecology* 50 (1969), 950–961.

26. J. A. Estes and J. F. Palmisano, "Sea Otters: Their Role in Structuring Nearshore Communities," *Science* 185 (1974), 1058–1060.

27. C. G. Jones, J. H. Lawton, and M. Shachak, "Organisms as Ecosystem Engineers," *Oikos* 69 (1994), 373–386.

28. M. E. Power, D. Tilman, J. A. Estes, B. A. Menge, W. J. Bond, L. S. Mills, G. Daily, J. C. Castilla, J. Lubchenco, and R. T. Paine, "Challenges in the Quest for Keystones: Identifying Keystone Species Is Difficult but Essential to Understanding How Loss of Species Will Affect Ecosystems," *Bioscience* 46 (1996), 609–620.

29. As noted by Jones et al., "Organisms as Ecosystem Engineers."

30. Darwin, *Origin of Species.*

31. The term "biodiversity" came from the National Forum on BioDiversity in Washington, D.C., held in 1986. E. O. Wilson, "Introduction," in M. L. Reaka-Kudla, D. E. Wilson, and E. O. Wilson, eds., *Biodiversity II: Understanding and Protecting Our Biological Resources* (Washington, D.C.: Joseph Henry Press, 1997), 1–3.

32. R. B. Waide, M. R. Willig, D. F. Steiner, G. Mittleback, L. Gough, I. Dodson, G. P. Juday, and R. Parmenter, "The Relationship between Productivity and Species Richness," *Annual Review of Ecology and Systematics* 30 (1999), 257–300.

33. D. M. Post, M. L. Pace, and N. C. Hairston, Jr., "Ecosystem Size Determines Food-Chain Length in Lake," *Nature* 405 (2000), 1047–1049; Rosenzweig, *Species Diversity.*

34. Note that the results here are based on laboratory studies. R. Kassen, A. Buckling, G. Bell, and P. B. Rainey, "Diversity Peaks at Intermediate Productivity in a Laboratory Microcosm," *Nature* 406 (2000), 508–512.

35. D. Tilman, D. Wedin, and J. Knops, "Productivity and Sustainability Influenced by Biodiversity in Grassland Ecosystems," *Nature* 379 (1996), 718–720.

36. Michael Huston quoted by J. Kaiser, "Rift over Biodiversity Divides Ecologists," *Science* 289 (2000), 1282–1283.

37. For example, see M. A. Huston, "Hidden Treatments in Ecological Experiments: Reevaluating the Ecosystem Function of Biodiversity," *Oceologia* 110 (1997), 449–460.

38. A. Hector, B. Schmid, C. Beierkuhnlein, M. C. Caldeira, M. Diemer, P. G. Dimitrakopoulos, J. A. Finn, H. Freitas, P. S. Giller, J. Good, R. Harris, P. Högberg, K. Huss-Danell, J. Joshi, A. Jumponen, C. Körner, P. W. Leadley, M. Loreau, A. Minns, C.P.H. Mulder, G. O'Donovan, S. J. Otway, J. S. Pereira, A. Prinz, D. J. Read, M. Scherer-Lorenzen, E.-D. Schulze, A.-S.D. Siamantziouras, E. M. Spehn, A. C. Terry, A. Y. Troumbis, F. I. Woodward, S. Yachi, and J. H. Lawton, "Plant Diversity and Productivity Experiments in European Grasslands," *Science* 285 (1999), 1123–1127; M. Loreau and A. Hector, "Partitioning Selection and Complementarity in Biodiversity Experiments," *Nature* 412 (2001), 72–76; D. Tilman, P. B. Reich, J. Knops, D. Wedin, T. Mielke, and C. Lehman, "Diversity and Productivity in a Long-Term Grassland Experiment," *Science* 294 (2001), 843–845.

39. M. Loreau, S. Naeem, P. Inchausti, J. Bengtsson, J. P. Grime, A. Hector, D. U. Hooper, M. A. Huston, D. Raffaelli, B. Schmid, D. Tilman, and D. A. Wardle, "Biodiversity and Ecosystem Functioning: Current Knowledge and Future Challenges," *Science* 294 (2001), 804–808.

40. L. Tangley, "High CO_2 Levels May Give Fast-Growing Trees an Edge," *Science* 292 (2001), 36–37.

41. P. B. Reich, J. Knops, D. Tilman, J. Craine, D. Ellsworth, M. Tjoelker, T. Lee, D. Wedin, S. Naeem, D. Bahauddin, G. Hendrey, S. Jose, K. Wrage, J. Goth, and W. Bengston, "Plant Diversity Enhances Ecosystem Responses to Elevated CO_2 and Nitrogen Deposition," *Nature* 410 (2001), 809–812.

42. In case you didn't guess, it was the plot with sixteen plants that "won."

43. Aspirin is a modified form of the salicin Hippocrates ground from willow bark and given to ease pain.

44. Bear root is *Ligusicum porteri*, also known as osha.

45. R. Andrews, "Western Science Learns from Native Culture," *The Scientist* 6 (1992), 6.

46. X. Valderrama, J. G. Robinson, A. B. Attygalle, and T. Eisner, "Seasonal Anointment with Millipedes in a Wild Primate: A Chemical Defense against Insects?" *Journal of Chemical Ecology* 26 (2000), 2781–2790.

47. Y. Zhu, H. Chen, J. Fan, Y. Wang, Y. Li, J. Chen, J. Fan, S. Yang, L. Hu, H. Leung, T. W. Mew, P. S. Teng, Z. Wang, and C. G. Mindt, "Genetic Diversity and Disease Control in Rice," *Nature* 406 (2000), 718–722.

48. J. W. Kirchner and A. Weil, "Delayed Biological Recovery from Extinctions throughout the Fossil Record," *Nature* 401 (2000), 177–180.

49. J. E. Richardson, F. M. Weitz, M. F. Fay, Q.C.B. Cronk, H. P. Linder, G. Reeves, and M. W. Chase, "Rapid and Recent Origin of Species Richness in the Cape Flora of South Africa," *Nature* 412 (2001), 181–183.

50. Wilson, *Diversity of Life*, 330.

51. J.-M. Claverie, "What If There Are Only 30,000 Human Genes?" *Science* 291 (2001), 1255–1257. There is every reason to believe that as we learn more, the known number of human genes will change.

9. Epilogue: The Keystone Species with a Choice

1. R. Bailey, "The Progress Explosion: Permanently Escaping the Malthusian Trap," in R. Bailey, ed., *Earth Report 2000* (New York: McGraw-Hill, 2000), 13.

2. S. Molnar and I. M. Molnar, *Environmental Change and Human Survival: Some Dimensions of Human Ecology* (Upper Saddle River, N.J.: Prentice Hall, 2000), 248.

3. These are rough numbers, but I tried to keep them reasonably conservative. There are over 130 million passenger cars in the United States, and I assumed that each day only 20 million would do excess driving to find a closer parking place while shopping or at work. I had each of them go three hundred yards (or idle for a minute or two while waiting for another car to pull out), at twenty miles per gallon. That results in over two million barrels of gasoline (petrol) per year. Based on statistics from the U.S. Energy Information Administration on each country's daily petroleum consumption, that results in 4.46 days in South Africa, 62.1 days in Zimbabwe, and 82.04 days in Georgia.

4. W. R. Catton, Jr., *Overshoot: The Ecological Basis of Revolutionary Change* (Urbana: University of Illinois Press, 1982).

5. C. Bright, "Anticipating Environmental 'Surprise,'" in L. R. Brown et al., eds., *State of the World 2000* (New York: W. W. Norton & Co., 2000), 22–38.

6. D. H. Meadows, D. L. Meadows, J. Randers, and W. W. Behrens III, *The Limits to Growth* (New York: Signet, 1972).

7. Lutz et al., "End of World Population Growth."

8. P. R. Ehrlich and A. H. Ehrlich, "The Population Explosion: Why We Should Care and What We Should Do about It," *Environmental Law* 27 (1997), 1187–1208.

9. Hardin, "Tragedy of the Commons," 1247.

10. One must be careful with such contributions — every day I get enough solicitations in the mail from conservation organizations that I wonder how many trees have been felled just to get the word out. There has to be a better way.

11. J. K. Smail, "Beyond Population Stabilization: The Case for Dramatically Reducing Global Human Numbers," *Politics and the Life Sciences* 16 (1997), 183.

12. Smail, "Beyond Population Stabilization," 190.

13. T. Dyson, "Better Focus on the Practical Challenges Posed by Population Growth (or, to Create a Forest You Must Plant Trees)," *Politics and the Life Sciences* 16 (1997), 195–197.

14. Cohen, *How Many People*.

15. Again I stole a gem of an expression from Paul Ehrlich's radio interview (see note 36 to chapter 4).

Index

About the Author

JEFFREY K. MCKEE is an associate professor of anthropology at The Ohio State University. After earning his Ph.D. from Washington University in 1985, he spent nearly a decade conducting research on human evolution and paleo-ecology in South Africa. He has published widely, including a coauthored book on human evolution and *The Riddled Chain*.